POLITICS OF SHORE EROSION: WESTHAMPTON BEACH

POLITICS OF SHORE EROSION: WESTHAMPTON BEACH

JOSEPH M. HEIKOFF

Professor of Public Administration
Graduate School of Public Affairs
State University of New York at Albany

ANN ARBOR SCIENCE
PUBLISHERS INC
P.O. BOX 1425 • ANN ARBOR, MICH. 48106

P.O. Box 1425, Ann Arbor, Michigan 48106

Library of Congress Catalog Card No. 76-1727
ISBN 0-250-40130-4

Manufactured in the United States of America

This research was sponsored by the New York Sea Grant Institute under a grant from the Office of Sea Grant, National Oceanographic and Atmospheric Administration (NOAA), U.S. Department of Commerce. The U.S. Government is authorized to produce and distribute reprints for governmental purposes notwithstanding any copyright notation appearing hereon.

New York Sea Grant Institute
Albany 1976

PREFACE

This investigation of a shore erosion problem at Westhampton Beach, Long Island had two objectives. One was to find out how scientists and engineers defined the problem and what they recommended as a course of action to deal with it. The other was to see how this information was used by the political decision-makers and governmental administrators at the federal, state, and local levels in determining policy and a program of action.

Perusal of some of the literature on beach erosion and technical reports prepared for the public agencies involved indicates that knowledge about the problem is incomplete and proposed solutions are controversial. Political decision-makers therefore had to try to reconcile or choose between a variety of confusing alternatives. The U.S. Army Corps of Engineers advocated completing the planned groin field along Westhampton Beach. Other engineers and professionals in the field opposed the construction of additional groins and proposed other courses of action.

Who was right? Whom can political leaders trust for the "right" technical solution to environmental problems? In the face of the ambiguities and uncertainties presented to them by engineering and science professionals, it is clear that politicians will do what comes naturally—follow their political instincts. This report on the Westhampton Beach case does not pretend to present either a definitive analysis of the erosion problem or "the" solution. It only demonstrates the difficulty of doing either and what this may mean for the efforts of governments to carry out their responsibility for managing coastal resources.

The Westhampton Beach case also indicates that intergovernmental relations may prove to be one of the most serious limiting factors on attempts to manage environmental resources in the coastal zone. Perhaps the most serious issue in these relations is the distribution of policy-making and regulatory powers between the state and local governments. Local and statewide interests may be very different regarding resources conservation versus development for income and tax producing enterprise, energy and fuel production, or regional recreation. It is not surprising, therefore, that resolution of this issue will involve considerable political struggle for the determining voice in decision-making.

This investigation would not have been started without the interest of Dr. Donald F. Squires, Director of the New York Sea Grant Institute. I am grateful to him and to his staff for continuous support and cooperation. Dr. Paul Marr, group leader for policy studies, was particularly helpful in providing counsel and project management. Dr. Orville Terry, science editor, provided needed quality control and valuable suggestions for improving and correcting the manuscript, as did Jean McAlpine, assistant director for communications.

Sea Grant Advisory Service member Roger Allbee made initial introductions to public officials, and Peter Sanko read the manuscript. I am also grateful to Dr. Donald R. Coates of SUNY at Binghamton, Dr. Lee E. Koppelman, Director of the Nassau-Suffolk Regional Planning Board, and DeWitt Davies of the Regional Marine Resources Council for reading and commenting on the manuscript. The manuscript was also read by Eldred Rich and members of his staff in the New York Department of Environmental Conservation. They made available valuable materials to document the study. Gilbert Nersesian, Chief, Beach Erosion and Hurricane Protection Section, New York District, provided access to Corps of Engineers documents and the photographs of Westhampton Beach. Donna Parker and Ada Bradley typed the manuscript and its many revisions.

For all of these contributions I am grateful, but I must take responsibility for the shortcomings of this final report.

Joseph M. Heikoff
Albany, New York
December 1975

For Sara and Barbara

CONTENTS

CHAPTER 1

INTRODUCTION

Coastal zone management is a complex activity in which all levels of government participate. The Coastal Zone Management Act of 1972 gives the federal government a new role and new responsibilities for defining a national coastal zone management policy and encouraging the states to formulate their own policies. The Act also authorizes the federal government to make financial support available to the states for planning and implementing measures for balanced development and conservation of coastal zone resources.

Although federal and state policy-making activities are the primary focus of the Coastal Zone Management Act, local governments have important powers and responsibilities for planning and managing the coastal zone within their jurisdictional boundaries. Within the guidelines of state enabling legislation, local governments wield powerful zoning and other land-use controls. Even when state and federal grants are available for planning and implementation of environmental programs, local governments must often take the initiative in recognizing local environmental problems and requesting state and federal participation in dealing with them. Local governments must also usually provide matching funds for project implementation. In any case, whether environmental development and conservation projects are controlled or funded by federal, state or local authorities, their final impact is on the local community. The local economy, property values, way of life, and enjoyment of local environmental resources are all affected by governmental decisions made at every level.

STUDY OBJECTIVES

New York State has undertaken the task of preparing a management plan for its coastal zone. Planning is a complex and difficult process fraught with uncertainty. Planners must try to identify the political, economic, and environmental factors that have led to the creation of our present man-made and man-changed environment, with its inextricable mix of benefits and problems. They must also try to explore alternative futures as the basis for formulating policy and action options that may be open to both the private and public sectors in the coastal zone. Unravelling the tangled web of cause and effect relationships that motivate and bring about development and change in the environment, forecasting how they may operate in the future, and proposing how they *should* operate to achieve defined social objectives is difficult enough; this, however, is only one aspect of the planning process. Policy decisions must be made and implemented, presumably in light of the information made available by the planners about action alternatives and their probable consequences, by elected political leaders at all governmental levels who are responsible for deciding the fate of the coastal zone.

As New York State enters upon its new coastal zone planning endeavors under the federal Coastal Zone Management Act of 1972, a reminder of past experience may be helpful. There has been a long history of federal, state, and local concern about the coastal zone and efforts to deal with its problems. This investigation attempts to identify the scientific and technical basis for planning decisions in one case of coastal zone management, and how this information influenced intergovernmental decision-making. The coastal zone management problem selected for study is beach erosion and hurricane protection at Westhampton Beach, one of the barrier islands on the Long Island south shore between Moriches and Shinnecock Inlets.

CONCEPTUAL FRAMEWORK

Planning and implementation for coastal zone management are products of the interaction of political and technical considerations. This case study attempts to identify the political and technical factors that were taken into account in the decision-making process for dealing with the Westhampton Beach erosion problem.

Coastal zone management itself has two aspects that might be seen as positive and negative in character. The positive aspect of management is developmental: it is concerned with the design and implementation of projects and programs to deal with particular problems of the coastal zone, such as protection of the shoreline from erosion, maintenance of navigation channels, preservation of wetlands, and control of pollution. The positive aspect of coastal zone management takes the form of direct government action. The negative aspect is regulatory rather than developmental: it is concerned with the guidance and limitation of private action in the coastal zone. Zoning and other land-use controls, limitation of air and water pollutant emissions, and permit procedures for construction, dredging, and other physical modifications of the coastal zone by private interests are examples of public regulatory measures.

The relationship of these four factors is indicated in the matrix shown in Figure 1. The diagram indicates how political

Aspects of Coastal Zone Management	Criteria for Decision-Making	
	Political Considerations	Technical Considerations
Positive:		
Public projects and action programs	Distribution of political and economic benefits and costs resulting from public action alternatives	Analysis of environmental problems in the coastal zone, identification of action alternatives, and design and implementation of projects
Negative:		
Regulation and control of private action	Distribution of political and economic benefits and costs resulting from regulation of private activities	Design of legal and administrative measures to regulate private action in the coastal zone

Figure 1. The decision-making process in coastal zone management.

and technical considerations interact to produce the developmental and regulatory aspects of coastal zone management. The objective of this case study has been to describe the direct action components of the decision-making process.

Government Planning and the Coastal Zone

Planning responsibilities are vested in state, regional, county, town and municipal organizations. State, county, and local objectives for the coastal zone are not always congruent, however, and the same powers to implement plans and regulate private development are not available to all governmental jurisdictions. Other institutional elements in the complex structure of overlapping and possibly conflicting planning concerns are single or multipurpose special districts and federal agencies.

Local governments are both aided and restricted in their attempts to manage the segments of the coastal zone within their jurisdiction. Intergovernmental technical and financial assistance programs for planning and development are available, but local authorities must also contend with neighboring local governments and state and federal agencies in defending or promoting their own interests. Coastal zone management on Long Island, therefore, is not guided by a single comprehensive planning program but is influenced instead by a multiplicity of local plans as well as efforts to coordinate them by regional, state, and federal agencies.

An important intergovernmental concern in coastal zone planning is the respective roles played by state and local governments in the planning process. Local governments appear to have primary responsibility for environmental and land use planning in their jurisdictions. They also exercise controls over land use and other development. What, then, is the role of the state in planning? How would a "state plan" relate to local plans? One approach to the answer may be found in the state's responsibility for intergovernmental planning coordination to assure that regional and local planning activities add up to a coherent state-wide plan for balanced development and protection of the state's resources. This study will examine how the state carried out this function in dealing with the Westhampton Beach erosion problem.

COASTAL ZONE MANAGEMENT PROBLEMS

The problems of the coastal zone in New York State may be defined and described in many different ways depending on the interests of the observer. Scientists, engineers, administrators, policy-makers, business managers, and property owners identify these problems in light of their own values, objectives and interests in the lands and waters of the coastal fringe. The Regional Marine Resources Council of the Nassau-Suffolk Regional Planning Board has identified 17 different coastal zone management problems.

1. Reduction of commercial shell fish production
2. Depletion of sport and commercial fin fisheries
3. Control of insects and related pests
4. Disposal of solid wastes
5. Destruction of wetlands
6. Development of marine industries
7. Stabilization and protection of the coastal shore
8. Dredging and spoil disposal
9. Eelgrass
10. Disposal of domestic wastes
11. Boat pollution
12. Oil spill pollution
13. Limited shoreline recreation facilities
14. Duck waste pollution
15. Salt water intrusion into fresh water supplies
16. Thermal pollution
17. Preservation of sites of natural or historic value.

These problems are all interrelated, for they are aspects of indivisible and complex geomorphic and ecological systems. A crude indication of some of the factors that influence changes in the physiographic structure and ecological systems in the coastal zone is given in Figure 2. The natural processes involved in sedimentation and erosion appear to be the principal factors that cause changes in the physical structure. Climate

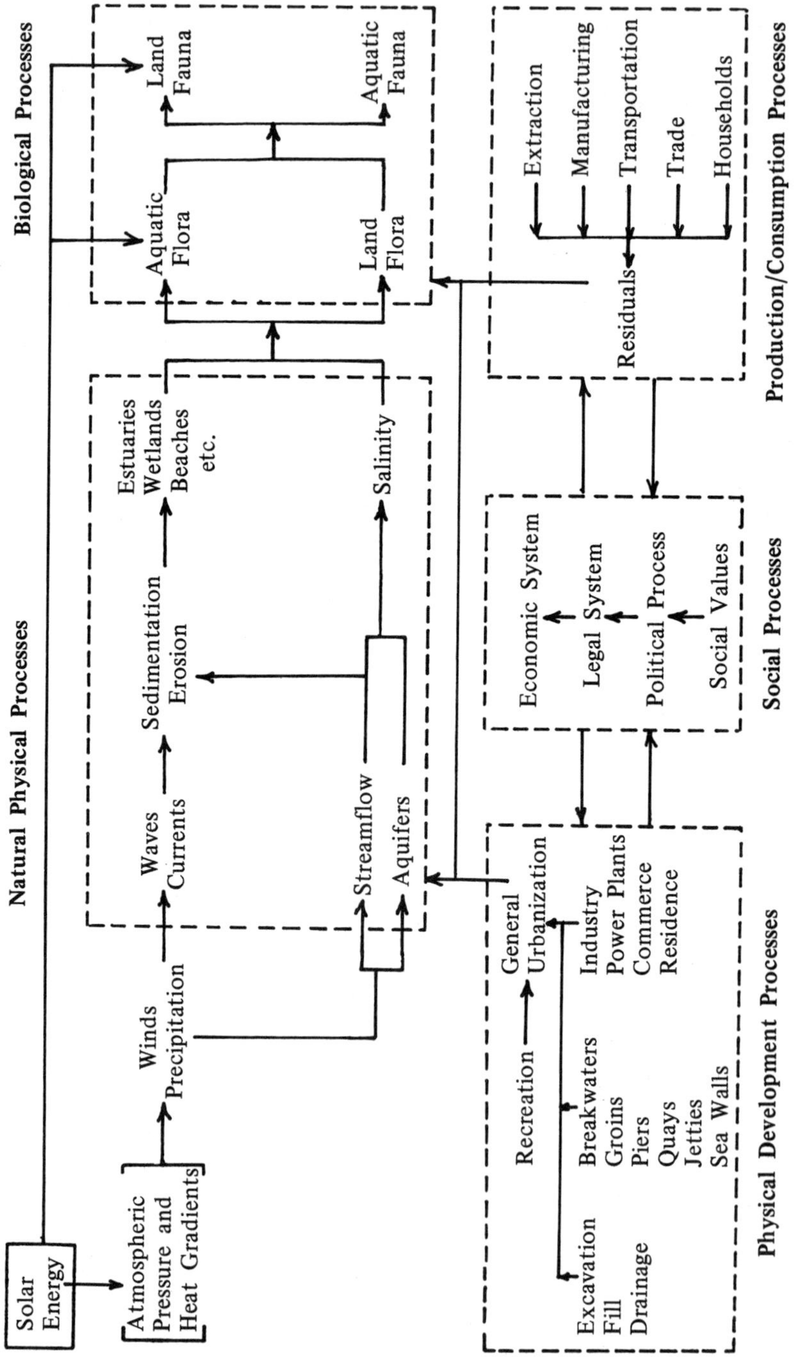

Figure 2. Change factors in the coastal zone.

and salinity changes appear to be major environmental factors influencing the ecosystems. Both physical structure and biota are also influenced by man-made physical development and production/consumption processes. These, in turn, are influenced by economic motivation, government regulation, political competition, and other social processes.

Action to deal with any one of these problems will affect others. Indeed, efforts to reduce one problem may exacerbate others. It is because men attach different values, economic and otherwise, to the biological and physical processes that take place in the coastal zone that they impute costs and benefits to the consequences of natural forces and human action. In the absence of human values, natural traumatic events and secular changes create and destroy geomorphic structures and plants and animals, but there is no one to reckon gain and loss. But when men possess, invest in, and exploit the land and water resources of the coastal zone, then they struggle with nature and with each other for individual advantage.

In the calculus of individual benefits and costs, the problems of coastal zone management appear to be reducible to the problem of public and private choice between letting nature take its course, creating an environment completely controlled by technology, or adopting policies somewhere between these polar extremes of natural anarchy and totalitarian human control. Nature cannot be natural with even a little control, and a completely artificial environment could be created only by single-minded social decision-making.

When economic and political democracy are preferred, attempts to maximize individual values cause a welter of conflicting actions in a market of private and social gains and losses. When private gain results in unbearable social loss, public authority asserts its mandate—sometimes too late—to protect the public interest. Here we come to the current governmental difficulties of defining and coping with the problems of coastal zone management. When is social loss out of balance with private gain? How can an event, action, or phenomenon be defined as a problem when the consequence is loss for some but gain for others? Hence the over-simplified reduction of the problems of coastal zone management to the basic public choice between conservation and protection of land and water resources versus their exploitation to maximize relatively short-term private gain.

Within this decision framework, coastal zone management is seen as basically a problem of politics to justify and define governmental constraints on private as well as public action. It is within this political context, confusing as it may be, that scientists and engineers must attempt to define technically the problems of the coastal zone. Unfortunately, these problems can have no "objective" reality, for they are infused with the "subjective" reality of the conflicting human interests that impinge on the coastal zone. It is the purpose of this investigation to explore that "subjective reality" as it is manifested by the various governmental and private interests that converge on the south shore of Long Island.

To avoid simplistic reductionism and at the same time divide the problems of the coastal zone into manageable components for case studies of intergovernmental decision-making, the four priority problems identified by the Regional Marine Resources Council have been adopted as the framework for this study. These problems are:

1. coast stabilization and protection
2. wetlands management
3. dredging and dredge spoil disposal
4. integrated water supply and waste water disposal[1]

The Westhampton Beach case study conducted under this project concerns the first of the four problems noted above. The issue confronting federal, state, and local decision-makers revolves around the question of whether or not six additional groins should be constructed to control erosion on Westhampton Beach east of Moriches Inlet.[2] The nature of the problem, the positions taken by the affected governments, and the decision-making process will be recounted in the following chapters.

FOOTNOTES

[1] Regional Marine Resources Council, *Guidelines for Long Island Coastal Zone Management*, Nassau-Suffolk Regional Planning Board (1973), pp. 2-3.

[2] Groins are long, narrow structures built of stone extending into the water at right angles to the beach.

CHAPTER 2

BEACH DYNAMICS AND EROSION

Beaches are relatively unstable land forms. They change from season to season and from year to year, and they are shaped by the natural forces of winds, waves, and tides, and by the works of man. Although seasonal changes and severe storms cause short-term radical alterations in beach forms, there is a cyclical pattern in these changes. Destruction and recovery alternate so that there is a natural equilibrium, which produces long-term changes in beach structure that are relatively small. This natural equilibrium cannot be maintained, however, when man takes possession of the beaches. His concept of real estate requires that land be considered an unalterable physical entity. A patch of sand that can be washed away, together with the structures on it, is not a saleable commodity. When men carve up the beach into lots, thereby legally if not actually converting it into real property, they wish, like King Canute, to stop the action of the waves and the tides.

WINDS, WAVES, AND SAND

Beach erosion, as a problem for the property owner (whether private or public), may be understood from a brief account of beach structure and dynamics.[1] If the property owner is dissatisfied with the way his segment of the beach behaves, he must blame the sun and the moon. The sun heats the earth and the air around it, but not uniformly. Warm air rises, causing an area of low atmospheric pressure. Colder air from areas of higher pressure moves into the low, thereby creating winds. Winds blowing over the surface of the ocean create waves, which strike the beach with more or less force, depending on the severity of the winds that generate the waves. Sun and moon together exert gravitational forces on the ocean, causing the tides.

The turbulent breaking waves that strike the shore stir up the beach materials and cause the sediments to be suspended and moved by the water for longer or shorter periods, depending on the size of the sand grains and the energy of the waves. Some of the sand moves at right angles to the beach; some of it moves in a saw-tooth pattern that results in the drift of sand parallel to the beach. The proportion of sand that moves perpendicular and parallel to the beach varies with the severity of the waves. During stormy periods, the perpendicular movement predominates, and sand is carried away from the beach, causing erosion. In calm periods, sand is carried back toward the shore, but there is also a greater proportion of sand movement parallel to the shore. This movement is caused by the littoral current, which may also cause erosion or accretion of the beach.

The succession of waves striking the beach at an angle sets up a longshore current that moves parallel to the shore in shallow water. This current carries the suspended sand grains in a littoral drift from one part of the beach to another. Because of the direction of the prevailing winds and waves, the littoral drift carries sand westward along the barrier beaches of the south shore of Long Island. This process maintains the western beaches, but only if there is a steady supply of sand from the east. If the supply to a particular section of beach ceases or is obstructed in some manner, then the turbulent waves and the littoral current will pick up sand already there and carry it westward, causing erosion of this beach segment.

The action of waves and littoral drift builds up a little ridge of sand along the high water line. This is the crest of the flat area of the beach, which is called the berm. Onshore winds pick up sand from the berm and carry it landward, building up the beach beyond high water and creating a dune ridge parallel to the shore line. A wide beach backed by a high dune is a most attractive location for houses and hotels. This becomes high value "real estate," although beach and dunes may have been washed away by the last hurricane or northeaster and may be damaged or destroyed by the next.

The cycle of destruction and recovery actually takes place on a comparatively small scale every year as the seasons change. Winter waves and storm surges wash sand from the beach into deeper water, and may even attack the face of the dune, forming a winter bar parallel to the shore. The moderate waves and swells of summer reverse this process. Summer waves are lower and have longer wave length. They pick up sand from the winter bar and return it to the beach face.

Waves and tides moving against the shore create surges and currents. Strong winds and high tides cause the water to pile up against the land, raising the water level by many feet, especially during severe storms. This causes flooding behind the beach as well as destruction of dunes and the beach itself by severe storm surges. Waves during calmer periods may eventually repair most damage to the beach by bringing sand from other areas, and the wind will rebuild the dunes, sometimes moving them a few or many feet landward in the process. Houses and other structures on the beach will have been destroyed, but they may be replaced after the beach and dunes have been reconstituted. This is costly, however, and the new landforms will not necessarily conform to the old property lines. The long-term movement has usually been landward over historical time, so that part of the beach simply disappears into the ocean. Hence the demand by property owners for some form of engineering works to stabilize the beaches and the land behind them in order to protect their investments.

BEACH CHANGES ON THE SOUTH SHORE OF LONG ISLAND

Most of the south shore of Long Island consists of relatively straight, narrow barrier islands separated from the irregular mainland shore by Great South Bay and other wide, shallow bays. Tidal currents exchange water between these bays and the Atlantic Ocean through narrow inlets between the barrier islands. These inlets are important features of the geomorphology or physical structure of the area, and play an important role in determining what happens to the beaches on the barrier islands.

Geomorphologic changes of East Coast barrier islands have been studied intensively by scientists, but there is not yet complete agreement among them about how these changes take place. In the absence of proven theories of beach dynamics, engineers are not able to predict to their own complete satisfaction the effect of engineering works on the future structure and stability of beaches where they have been installed. Nevertheless, public demand for shoreline stabilization and protection against hurricane damage, especially by owners of waterfront property, has forced governments to take action. Some of the scientific studies and engineering design criteria that provided technical inputs for decisions about the Westhampton Beach erosion problem are described in this book.

The beach erosion problem on Westhampton Beach between Moriches Inlet and Shinnecock Inlet, which is the focus of this investigation, is perhaps most strongly influenced by the movement of water and sand through these inlets. The location map, Figure 3, indicates the relative positions of the coastal features noted in the following account. This is based on an analysis by C. Larry McCormick, who is at Southampton College of Long Island University.[2]

The barrier beaches along the south shore of Long Island receive much of their sand from the erosion of cliffs up to 60 feet high in the Montauk area at the eastern end of Long Island. The glacial deposits that form these cliffs contain boulders, cobbles, gravel, and coarse and fine sand. The primary source of material for the barrier beaches is this sand, which is picked up by the prevailing westerly littoral drift. A secondary source is probably the larger material on the narrow beaches fronting the Montauk Cliffs. "The well-rounded nature of the cobbles and boulders on the beach indicates that the grinding action of these particles in the surf zone is actively manufacturing finer, more mobile constituents that may nourish beaches to the west."[3] It is estimated that the Montauk

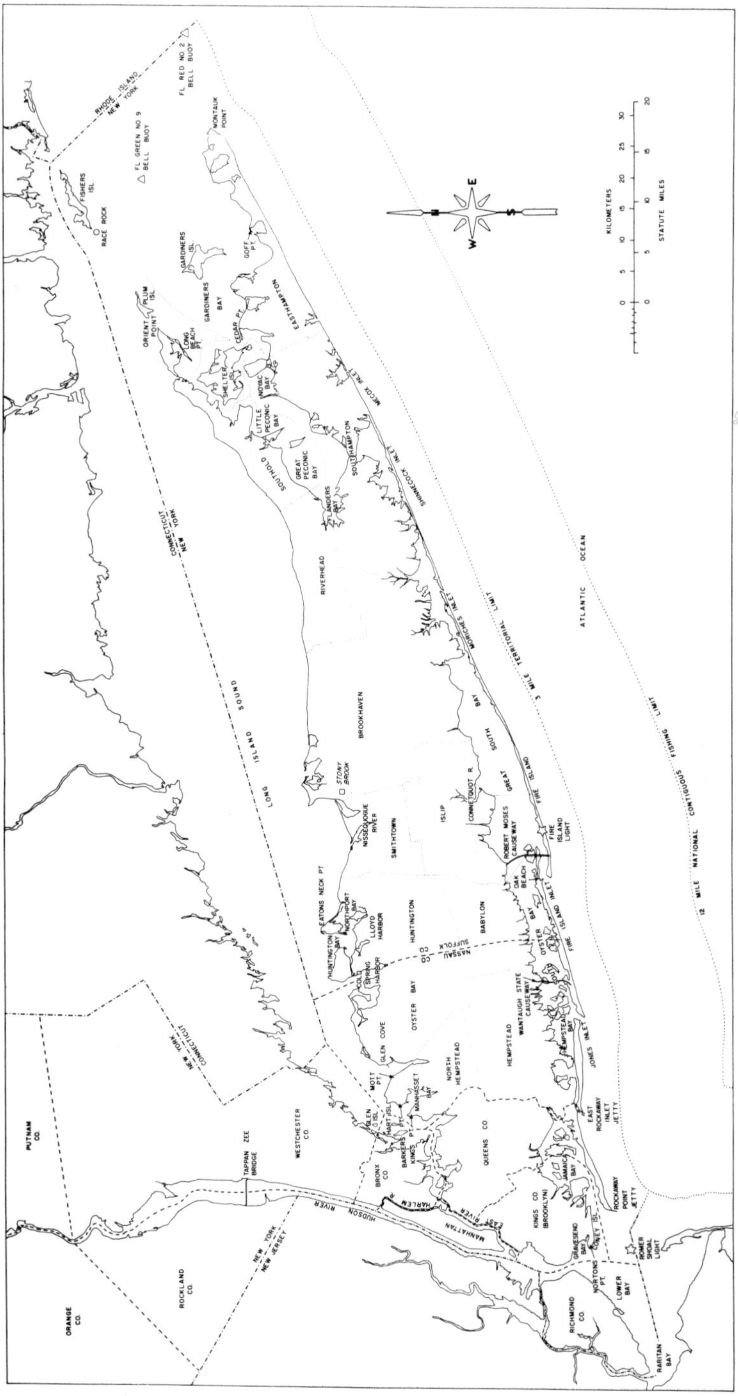

Figure 3. Location map.

Cliffs are receding at the rate of about one foot per year as the result of this erosion.

The westerly drift of sand from the Montauk Cliffs created a barrier island system that was continuous from the south shore mainland near Southampton to Fire Island Inlet, about 60 miles to the west. Fire Island Inlet is the first "permanent" natural break in the barrier island west of Montauk Point. This opening between the Atlantic Ocean and the Great South Bay has migrated westward, because the drifting sand from the east is constantly deposited at the western end of Fire Island, tending to close off the inlet from that side.

Since 1938 storms have broken through this reach of the barrier islands, but the natural processes described earlier eventually repaired most of the breaks by the drifting of sand from the east. However, "Moriches Inlet opened in 1931, and except for a one-year period, has remained open to the present. If it were not for attempts to stabilize this inlet in 1947 and 1952 it would probably have closed. Shinnecock Inlet (Figure 4) has a similar history; it was opened in the hurricane of 1938 and by 1952 was stabilized by the construction of stone jetties."[4] These artificially maintained inlets are probably the major causes of beach erosion on Westhampton Beach and Fire Island. The explanation of this situation lies in the interruption by the inlets of the westerly littoral drift of sand from Montauk.

The United States Coast and Geodetic Survey and the Army Corps of Engineers surveyed the shoreline in 1838, 1891, 1933, 1940, and 1956. During this 118-year period the beach east of Shinnecock Inlet to East Hampton retreated within the relatively narrow limits of 0.9 ft/yr to 1.4 ft/yr. During the first 95 years of this period, from 1838 to 1933, the erosion rate for the segment of the beach from Shinnecock to Moriches Inlets (Westhampton Beach) also was in the moderate range of 1.6 ft/yr to 0.7 ft/yr. From 1933 to 1956, however, "the Shinnecock to Moriches portion of the beach exhibited a dramatically increased rate of 6.8 ft/yr."[5] These annual erosion rates do not convey the magnitude of the retreat of the beach. On the average, Westhampton Beach receded 272.5 feet between 1838 and 1956, but the loss was over 500 feet in some places.

Figure 4. Shinnecock Inlet opening toward the Atlantic Ocean, showing the stone jetties on either side of the inlet that were constructed to keep it open.

The cause of this rapid beach erosion since 1933 is presumed to be the action of the new, artificially maintained inlets, trapping sand from the littoral drift. As the sand carried by the shore currents, usually westward but sometimes in the opposite direction, passes the inlets, some of it is picked up by the tides. The flood tide carries the sand through the inlets into the bay and deposits it in the form of a delta on the bay side of the barrier island. The ebb tide forms a similar, but smaller, delta on the ocean side of the inlet. The sediment trapped by this process and deposited in the flood tidal delta probably explains the accelerated erosion rates at Westhampton Beach since the Shinnecock Inlet was opened. McCormick cites the following calculations:

> To find if this process does remove a significant amount of sand from the littoral drift, the volume of sand being trapped was calculated from aerial photographs of the flood tidal delta taken in 1955 and 1969. The total volume added was found to be 22.3 million cubic feet, or 1.6 million cubic feet per year.[6]

> The volume of sand lost annually from beaches between Moriches and Shinnecock Inlets was 0.8 million cubic feet during the interval from 1838 to 1933. From 1933 until 1956 the annual loss of sand from the beaches between Moriches and Shinnecock Inlets increased to 4.6 million cubic feet, a difference of 3.8 million cubic feet per year. A minimum loss of 1.6 million cubic feet per year is accounted for by trapping sand on the flood tidal delta at Shinnecock Inlet. If it is assumed that Moriches Inlet, which is similar in size and geometry to Shinnecock Inlet, has the same annual trapping effect, nearly all of the 3.8 million cubic feet per year can be accounted for by loss to the inlets.[7] [See Figure 5]

Even before these inlets were opened there was a steady, but much slower, erosion of the beach. Nevertheless, the narrow barrier island did not disappear. Retreat of the beach was compensated by accretion on the bay side. Onshore winds carry sand from the beach berm into the dunes. During storms, waves may carry sand from the dunes to form washover fans of sediment on the other side of the island. Thus wind and wave action combine to produce a slow, natural landward migration of the barrier island. This process has been disturbed, however, by the severe loss of sand to the inlets. In recent years the beach has eroded faster than the buildup of new land on the bay. As the barrier island becomes narrower, the threat of washovers and breakthrough of new inlets by severe storms increases. The consequent damage to property on the island is obvious, but also important would be the increased vulnerability of the mainland shore on the bay to storm flooding. Change in salinity of the bay would decrease shellfish yields; and changes in bay currents could close navigation channels and reduce the flushing of pollutants from the bay into the ocean.

The Erosion Problem at Westhampton Beach

Governmental response to the beach erosion problem between Moriches Inlet and Shinnecock Inlet was the construction of a groin field along part of this reach of the barrier island (see Figure 6). The groin field did not include the westerly portion of the beach within the jurisdiction of the town of Southampton, however. This area continued to suffer from erosion, and in the early spring of 1973 a severe storm washed over the barrier

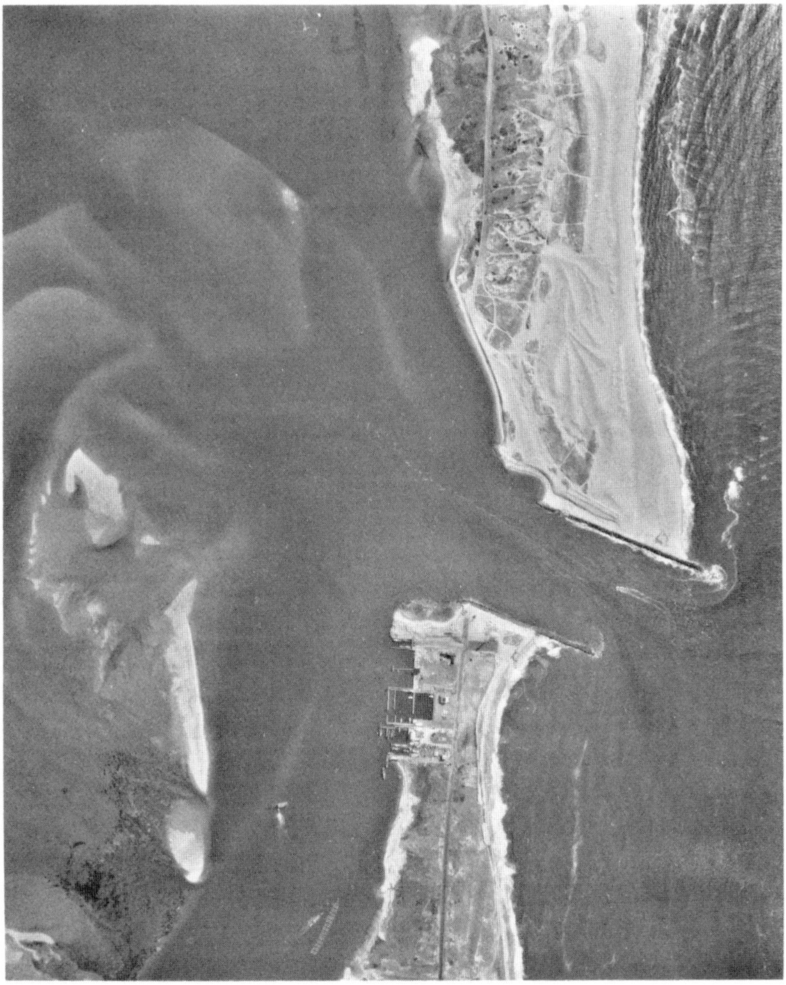

Figure 5. Shinnecock Inlet opening toward Shinnecock Bay. The tidal delta in the bay formed by the trapping of sand from the littoral current appears as a long narrow bar opposite the mouth of the inlet.

Figure 6. Stone groin at Westhampton Beach.

island in this location, washing out the town beach and parking lot and damaging adjacent private properties (see Figure 7). This immediate crisis intensified a political issue that had engaged the area for years. This was whether or not to extend the groin field westward toward Moriches Inlet and rebuild the beach.

The studies and events that led to the construction of the existing groins will be described in the following section. It may be noted here, however, that the issue revolves around such questions as: Have the existing groins aggravated the erosion problem at the western end of the barrier beach? Would building additional groins solve the problem? Should public funds be used to protect the value of private property on the beach?

The Effect of Groin Fields

A single groin interrupts the transport of sand along the coast, causing accretion of the beach on the updrift side and

Figure 7. Westhampton Beach near the Southampton Town Neighborhood Recreation Center, which was severely eroded in 1973. This section of beach is west of the existing groin field.

erosion on the downdrift side. Sand being transported by the shore current is trapped on the updrift side of the structure. The beach on the downdrift side is thus deprived of nourishment and at the same time is vulnerable to erosion by the current, which now has a reduced load of sand (Figure 8). If the erosion is severe enough, a second groin may be needed to protect this area. But the second groin causes erosion further down the beach, and a third one is required.

> Thus the construction of the first structure can set off a chain reaction that results in a requirement that the entire coastline be fronted by protective structures [Figure 9].
>
> The real "need" for a second or third structure may have been only temporary and may have diminished once the normal rate of longshore transport had been reestablished around the structure that caused the problem initially. However, if additional structures are built, the downcoast erosion becomes more severe with each succeeding structure, until finally a "point of no return" is reached where the need for additional protection from erosion

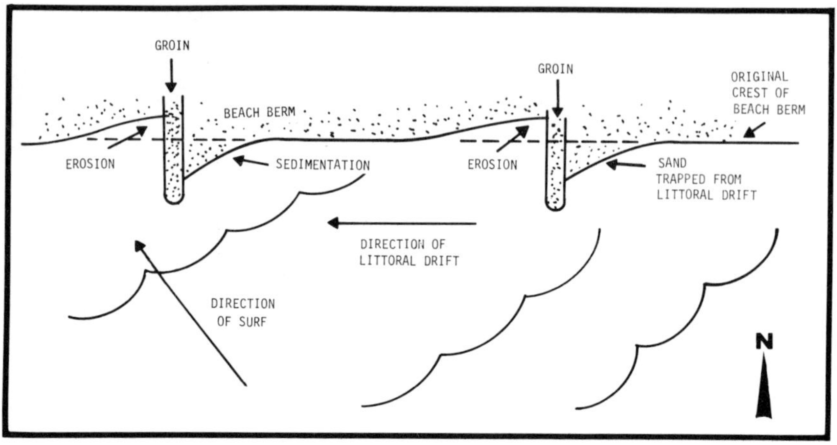

Figure 8. Effect of groins on beach.

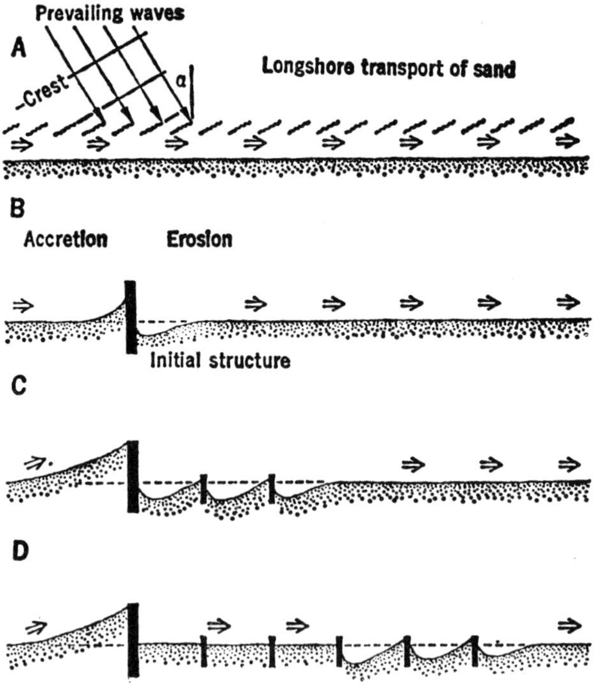

Figure 9. Erosional chain reaction following the installation of a single coastal obstruction: (A) straight beach with prevailing waves producing a longshore transport of sand, (B) accretion and erosion caused by initial structure, (C) downcoast erosion requiring two additional groins, (D) continuation of the downcoast erosion. Note that the first three groins are now unnecessary (after Inman and Brush, 1973).

> becomes so urgent that the only choices are (1) to continue to build protective works, (2) to find a new source of beach sand, or (3) possibly a combination of both.[8]

This appears to be the situation that has developed at Westhampton Beach. Each successive installation of a groin field—first 11 groins, then 4—has created public demand for more groins to the west, to rebuild the beach and to reestablish the dunes. The technical question to be answered is whether or not six additional groins are justified to complete the series originally designed by the Corps of Engineers. In seeking answers to this question, it is important to take a systems approach to analyzing the wind, wave, and sand movements at the shore and the total effect on these systems of man-made obstructions:

> Any guidelines fundamental to coastal planning necessarily involve an understanding of the natural physical processes active in the environment. It has been shown that these processes are driven by primary interrelated forces that are essentially systematic and regular in form. This observation is borne out by recent findings that the extent of the mixing of water in the surf zone can be predicted from a knowledge of the breaker height and the width of the surf zone, and the longshore transport of sand is directly proportional to the longshore component of the energy flux of the incident waves. Thus a coherent approach to the formulation of unifying planning criteria seems clearly within our grasp; and it is becoming apparent that the driving forces (that is, the nearshore "climate" which consists of waves, winds, and current in nearshore waters) and the budget of sediment in a given coastal region are the principal environmental measurements that must be assessed for effective planning. . . . Certainly, the future of any coastal man-made structure placed in the path of the longshore movement . . . is questionable, and great reservations should accompany any commitment to build such a structure. Aerial photographs of Miami Beach, Florida, or Cape May, New Jersey, show the abuse to the coastline that can result when cascades of groins are erected without a master plan.[9]

The Moriches to Shinnecock Inlet beach erosion control and hurricane protection project designed by the Corps of Engineers is such a master plan. Nevertheless, its execution in stages separated by long time intervals did not create an integrated system of rebuilt beaches and dunes and stabilizing structures.

Partial construction to protect some sections of this great beach seems to have accelerated erosion elsewhere.

CONFLICTING THEORIES OF BARRIER BEACH MANAGEMENT

The Corps of Engineers' strategy for dealing with beach erosion on the Long Island barrier islands included reconstruction of eroded beaches and dunes by pumping sand from the back bays and stabilizing the fill with stone groins. Other scientists who have studied barrier islands maintain that this strategy will contribute to further deterioration of the beach rather than help to preserve it. Their studies indicate that letting nature take its course without interference is a preferable strategy, because natural forces tend to maintain a barrier beach in a state of equilibrium that results only in slow changes.

Paul J. Godfrey and Melinda M. Godfrey have summarized studies carried out on the Outer Banks of North Carolina.[10] The conclusions about coastal management strategies reached by these studies were derived from comparison between those barrier islands in the Outer Banks that have been developed with intensive building, roads, and artificial high protective dunes and those islands that have remained in a predominantly natural condition. In both situations, there is an intimate relationship between the physical structure of the barrier islands and the plant ecology. The Godfreys and other scientists maintain that the natural geomorphic/ecological systems are in equilibrium with the very high energy winds and waves to which they are exposed, while the artificial systems are self-destructive.

The Natural Barrier Island System

As in the barrier islands of the Long Island south shore, the Outer Banks are influenced by two major natural forces: the secular transport of sand by the littoral drift, and the drastic short-term effect of severe storms. Core Banks is an island in its natural state. It has a low profile, so that storms overwash the island, without causing serious damage. The beach has a very wide berm. Behind the berm is a zone of irregular low dunes. Behind the dunes are overwash terraces formed by storm-surge deposition of sand, and farther back, on the sound side of the island, are salt marshes. This low island maintains a dense cover of sea oats (*Uniola paniculata*) and salt meadow cordgrass (*Spartina patens*) on and behind the dunes. The effect of storms on Core Banks island is described as follows:

> Storms maintain the wide, bare berm as waves dissipate their energy on the back shore and pass across the island. The berm is the "elbow room" for the storm surge. Whatever dunes or vegetation build up on the active berm during quiet periods are usually destroyed by the surge. Those dunes that survive are back far enough to escape the main wave force. As the overwash occurs, sand is carried from the beach and berm crest onto and across the berm to the interior of the island, where it is deposited as a veneer of sand in the form of a terrace or fan. Winds then blow the fine sand from the deposit into dunes. . . . As sand is moved into the interior of the island by overwash, and on dunes by wind, the height of the land increases and tends to keep pace with the rising sea. . . . Thus, much of the island consists of a series of quite recent overwash terraces. New dunes form on the terraces only to be buried or eroded by overwash in their turn. The sand, however, is carried farther back into the island during an overwash instead of being lost to the littoral currents, and so it is conserved.[11]

Instead of destroying the natural Outer Banks islands, periodic overwash maintains them and even builds them up. They are aided in this process by the native grasses. The grasses help to build the dunes as wind-blown sand accumulates around their stems. Both the plants and dunes grow vertically. *Spartina* will even come up after burial by overwash and grow more vigorously than ever. When overwash completely crosses the island, the new material is colonized by marsh grass, *Spartina alterniflora*, which is more luxuriant than in the older marshes.[12]

The Developed Islands of the Outer Banks

Development of Hatteras and other islands has led to the attempt at total stabilization of naturally dynamic systems. To prevent the sea from washing over the islands and to protect roads and buildings, it is necessary to hold all the sand in place, preventing dune movement and the opening of inlets. A continuous, high artificial dune line has been built. This is essentially a dike. The beach in front of the dune has become narrow, so that storm waves strike the dune instead of dissipating their energy on a wide berm. Since they cannot roll over the island, the waves rebound from the dune face, "eroding beach and dune sand into the littoral drift and out to deep water. . . . A dune that cannot retreat can only erode."[13]

The trapped energy of the waves erodes the beach in front of the dune; there is no overwash material, so the islands are subject to erosion from both sides, sound and ocean. Overwash is necessary to move sediments to the back side of the island, to protect the salt marsh fringe there, and to maintain the width of the island. Robert Dolan notes that "the artificial barrier dune system is associated with a change in beach width. The unaltered parts have beaches ranging from 125 to 200 meters wide. . . . Among many of the Hatteras and Bodie Islands beaches . . . the shoreline has receded to widths of 30 meters or less."[14]

An artificial plant ecology has been established on the artificial dune line. It has been planted with American beach grass (*Ammophila breviligulata*), which must be sprayed with fertilizers and pesticides.

> The effect of the barrier dune on the vegetation behind it has been to accelerate ecological succession in the direction of species less tolerant of flooding and overwash burial. Shrubs and other plants which would ordinarily be expected in the most sheltered parts of the island are now coming up very close to the sea, along with normal dune plants. A storm severe enough to breach the dune would reverse this trend, eliminating all but the most tolerant species.[15]

Coastal Management Strategies for Barrier Islands

The consequences of the different management strategies employed on the Outer Banks can be seen in the effects of Hurricane Ginger, which struck the islands on September 30, 1971. All along the developed beaches there was severe erosion. The seaward face of the barrier dunes near Cape Hatteras receded between 3 and 5 meters, on the average. The undeveloped islands were almost completely under water, but there was no permanent damage. The beach berm was flattened, but there was overwash buildup by which new sand was added to the backshore and the level of the interior was raised. The grasses recovered from burial by the overwash, and the beach was built seaward again by the normal waves and currents. In only 10 months, the beach had built out beyond its position before the hurricane struck.[16]

The Outer Banks islands that have been left alone are stable geomorphic features that maintain themselves in spite of periodic assaults by severe storms. In contrast,

> the Cape Hatteras sections show a narrow and irregular berm, and continuous artificial dune lines that are eroding in numerous places on every island. Some of this erosion is serious enough to have prompted massive and expensive control projects, usually in front of some threatened facility. Nearly $0.5 million was spent during the early part of 1972 for beach nourishment at Buxton, only to have it all disappear 10 months later. A $2 million project has been approved and is now in the planning stages to restore this same section of the beach. In spite of endless expensive erosion control efforts (between 1957 and 1968 the Park Service spent an average of $420,000 per year on dunes, with additional sums of $1,200,000 to repair storm damage), buildings and roads close to the barrier dune are constantly vulnerable to destruction by the sea.[17]

Over 40 years of National Park Service experience on the Outer Banks has led the agency to conclude that the enormous costs of trying to fight the sea were not worth the benefits derived. It announced in September 1973 that it was revising its policies. "After spending $21 million since the 1930s to erect and maintain artificial barriers against waves and storms, the agency has concluded that such work does more harm than good."[18]

Experience on the Outer Banks may indeed suggest that barrier islands should not be developed, and that no attempt should be made to create artificially high dunes. But how can this management strategy be applied to the Long Island barrier beaches? They are already intensely developed with many millions of dollars' worth of resort hotels, expensive private residences, public parks, and streets and utilities. It is no longer possible politically or economically to permit Westhampton

Beach to return to its natural state and permit violent storm surges to wash over it.

There is also an important difference between the development strategy used by the National Park Service at Hatteras and by the Corps of Engineers on Long Island. The Park Service program relied on a high continuous dune and exotic vegetation to stabilize the barrier islands. Groin fields were not used to stabilize the rebuilt beaches, as they were at Westhampton. Where the groins were constructed, the Westhampton beach has maintained itself, although the town beach area to the west has suffered. This problem was apparently caused, however, by failure to complete the Moriches to Shinnecock project as designed rather than failure of the design itself.

FOOTNOTES

[1] Detailed accounts of beach dynamics may be found in: U.S. Department of the Army, Corps of Engineers, *Shore Protection Guidelines*, August 1971, pp. 13-24, and F. L. Bartholomew and W. V. McGuinness, Jr., *Coast Stabilization and Protection on Long Island*, Regional Marine Resources Council, Nassau-Suffolk Regional Planning Board, February 1972, pp. 8-23.

[2] McCormick, C. Larry. "Probable Causes of Shoreline Recession and Advance on the South Shore of Eastern Long Island," in *Coastal Geomorphology*, D. R. Coates, Ed. (Binghamton, New York: Publications in Geomorphology, State University of New York at Binghamton, 1973), pp. 61-71.

[3] *Ibid.*, p. 62.

[4] *Ibid.*, p. 67.

[5] *Ibid.*

[6] *Ibid.*

[7] *Ibid.*, p. 69.

[8] Inman, Douglas L. and Birchard M. Brush. "The Coastal Challenge," *Science* **181**, 29 (1973).

[9] *Ibid.*, p. 30.

[10] Godfrey, Paul J. and Melinda M. Godfrey. "Comparison of Ecological and Geomorphic Interactions between Altered and Unaltered Barrier Island Systems in North Carolina," in *Coastal Geomorphology*, D. R. Coates, Ed. (Binghamton, New York: Publications in Geomorphology, State University of New York at Binghamton, 1973), pp. 239-257.

[11] *Ibid.*, pp. 243-245.

[12] *Ibid.*, pp. 245-248.

[13] *Ibid.*, p. 249.

[14] Dolan, Robert. "Barrier Islands: Natural and Controlled," in *Coastal Geomorphology*, D. R. Coates, Ed. (Binghamton, New York: Publications in Geomorphology, State University of New York at Binghamton, 1973), p. 267.

[15] Godfrey and Godfrey, p. 251.

[16] Dolan, pp. 271-274.

[17] Godfrey and Godfrey, p. 240.

[18] *New York Times* article datelined September 24, 1973.

CHAPTER 3

THE FEDERAL ROLE IN SHORE PROTECTION

Barrier beach erosion is the concern of all levels of government. Local governments are, perhaps, most immediately concerned, for the consequences of beach erosion affect the lives of their citizens, public and private property, and the community treasury. Nevertheless, local government resources are generally inadequate for either complete technical analysis of the problem or implementation of engineering or other works to cope with it. Beach erosion is also a problem that affects the entire coastline, which includes many communities. Control of beach erosion requires measures to deal with the entire area affected. Local attempts to control erosion by constructing such works as groins or jetties in the section of beach within the local jurisdiction may be ineffectual in the long run and may even accelerate damage to the local beaches and to those in other communities that depend on the littoral drift for beach nourishment.

The counties, regional organizations, and the state are therefore also directly concerned with this problem. Although they have larger geographic areas of concern and greater technical and financial resources, they are not in a position to assume full responsibility for dealing with barrier beach erosion. The federal government has recognized that the coastal zone is a national resource, and that it must also assume responsibility for dealing with this problem.

RECENT HISTORY

The recent history of federal concern for the nation's coastal resources goes back to 1930. Congress in that year authorized the Corps of Engineers to carry out studies for shore protection measures in cooperation with state and local governments. Federal participation in the construction of shore protection works was limited, however, to federally-owned shore front. In 1946 amendments to the law permitted the federal government to participate in construction projects where there were state and local publicly owned shores. In 1956 further amendments authorized federal participation in projects to protect private property where this was incidental to protecting public property or gaining other public benefits.

In 1946 a study of the Long Island south shore from Jones Inlet to Montauk Point was completed by the Beach Erosion Board, a former unit in the Corps of Engineers, in cooperation with the Long Island State Park Commission and Suffolk County. The study indicated a definite need for various forms of shore protection, but at that time the federal government was not authorized to participate in such projects. After the federal legislation was amended, however, further detailed studies were carried out in preparation for eventual construction. One of these was a study of the shore from Fire Island Inlet to Montauk Point. This study included consideration of the need for hurricane protection as authorized by Congress in response to severe storms that damaged the area in 1954 and 1955.

The Survey Report was published in 1958 and provided the basis for the projects constructed at Westhampton Beach. By 1966, eleven groins had been constructed in response to a disaster declaration by the President after the barrier island had been breached by a severe storm in 1962. Further storm damage and beach erosion to the west of this first groin field spurred the construction of four additional groins and the placement of dune and beach fill. This did not complete the shore protection project originally recommended in the 1958 Survey Report, however. The area to the west of the second groin field has experienced further erosion and a storm overwash in 1972.

The Westhampton Beach beach erosion and hurricane protection project was not completed as designed because the local governments decided not to provide their share of the required construction funds. The apparent local attitudes were described by Gilbert K. Nersesian, Chief of the Beach Erosion and Hurricane Protection Section, New York District, Corps of Engineers:

> What are some of the reasons for the lack of construction activities on these projects? Well, the Federal Government has, through your Congressional representatives, appropriated funds for construction of these projects. However, oftentimes the funds have had to be returned because of the lack of local cooperation which includes furnishing the local share of the cost. Also, looking in depth at some of the local attitudes, we find that people are somewhat complacent about the need for these projects. You can understand this attitude when you look back, and you find that it has been eleven years since the last severe storm, which was in March 1962, and about thirteen years since the last severe hurricane, which was Hurricane Donna in September 1960. So, with time, the memories of the devastation have passed and, also, we have had many people coming to live in these areas who were not aware of what happened there previously. In addition, there are local disagreements as to the support of these projects, and these disagreements are, oftentimes, along sectional lines. For example, people along the northern shore and central portions of the Island feel that they should not be paying for protection for the "rich people" along the barrier beaches or for those people who are subject to damages from these storms. "After all," they say, "they chose to go and build there, and enjoy their lives on these islands. Why should we pay for it?"
>
> Well, there is a public interest which has to be recognized. Many people say, "Oh, the Federal Government is spending millions at Westhampton Beach and I can't even get on the beach there." The project there is for combined beach erosion control and hurricane protection, but in the beach erosion control portion of that project, there is very little Federal cost sharing, based on the private ownership and use of most of this shore. The greatest portion of the Federal cost comes about because of the hurricane protection which derives broader benefits. That protection is eventually going to protect not only the people on the barrier islands, but it is going to protect the bay resources which includes the shellfish industry, and the communities along the mainland shore of the Island. This aspect is what has to be recognized. Also, you can't just build part of the project in one area and say, "This looks pretty

> good. Let's wait and see how this is going to do before anything else." Once you initiate a program, it has to be carried out until a viable portion of the project is complete; otherwise, you will never get the complete protection that is really needed. These are questions that still have to be resolved on a local level.[1]

Procedures for Federal Participation in Coastal Zone Projects

Engineering works to combat beach erosion are classed by the federal government as river and harbor projects. They require local initiation and financial participation, but they are authorized by Congress and designed and constructed under supervision of the Army Corps of Engineers. Procedures for federal participation are complex and time-consuming, requiring two separate acts of Congress and review by the Office of Management and Budget as well as other federal agencies.

The beach erosion control and hurricane protection project for the Moriches to Shinnecock barrier beach had to go through this long series of federal procedures beginning in 1954. Two stages of the project have been completed by beach and dune rebuilding and the construction of 11 groins in stage one and four groins in stage two. The proposal to continue beach rebuilding westward and add six more groins would constitute stage three of the project.

The tortuous route a project proposal must follow as it progresses through engineering survey, enactment in a federal river and harbor act, funding in an appropriations act, preparation of plans and specifications, and finally to construction is shown in Figure 10. The locally initiated proposal is sponsored by the area representatives in Congress, where it must first go through the House Public Works Committee for incorporation in a river and harbor bill. This committee approves the project only after detailed engineering surveys, public hearings, review by the governors of affected states and by affected federal agencies, and approval by the Secretary of the Army. If the project survives incorporation in a river and harbor act, it is authorized for construction; but no money is available for it. The Corps of Engineers must then shepherd it through review by the Office of Management and Budget and the

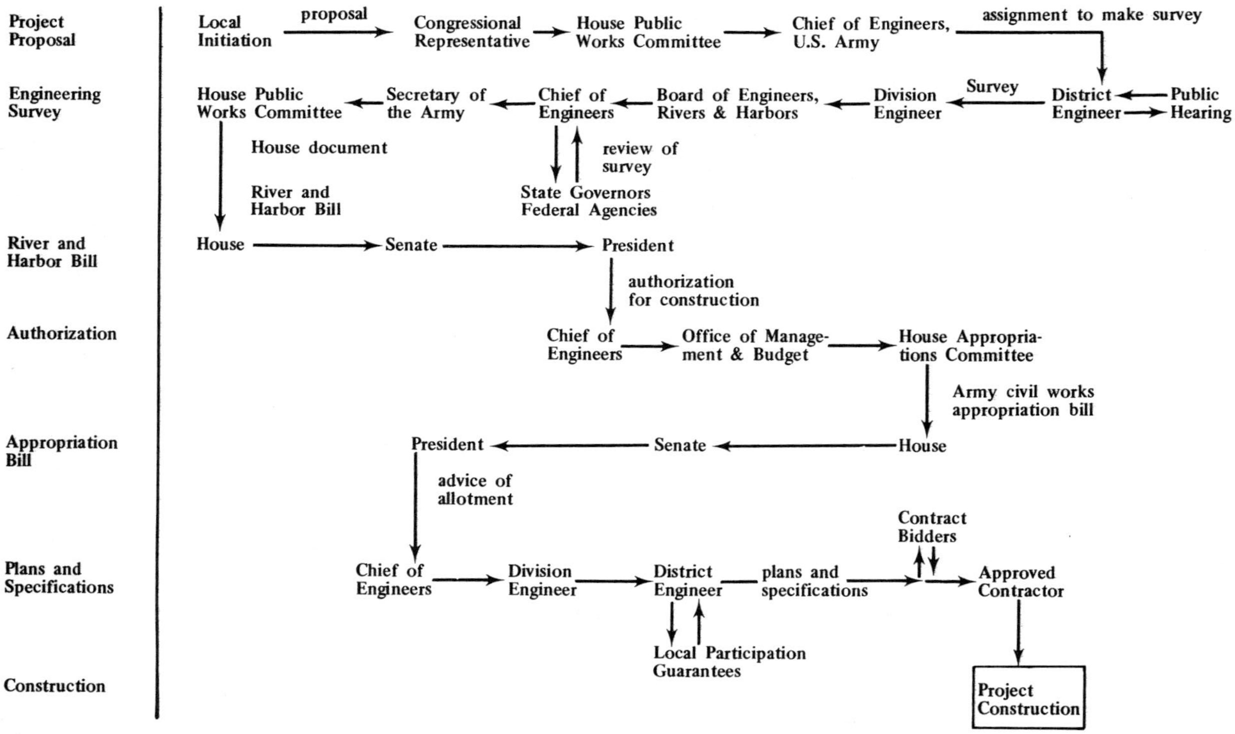

Figure 10. Procedures for federal participation in coastal zone projects.

House Appropriations Committee. Only after the project has been included in an appropriations act signed by the President will plans and specifications be prepared for eventual contract bidding and construction. During these stages the project passes up and down the bureaucratic hierarchy of the Corps of Engineers. Technical responsibility for engineering survey, design, and construction is in the hands of the District Engineer. The Long Island coastal zone is in the New York District of the Corps of Engineers, which has headquarters in New York City.

Not all projects must be specifically authorized by Congress, however. Small projects are defined as those for which the federal share of construction costs will be not more than $1 million. For such projects the Secretary of the Army can authorize a beach erosion study simply on request from a local agency. If the study shows the project to be justified, and the local interests involved are willing and able to cooperate in financing and providing other assistance, the Secretary of the Army can authorize construction and allocate funds for it from available civil works appropriations.[2]

Local Cooperation

State and local interests in shore protection and beach erosion projects are represented by the State Department of Environmental Conservation. During the years when the Moriches to Shinnecock Inlets project was surveyed, designed, and constructed, the State Department of Public Works and later the Conservation Department were the cooperating agencies.

The amount of federal participation in the financing of project surveys and construction depends on the public benefit derived.

> Under this concept, Federal participation is greatest where the protected shore areas are publicly owned and appropriate facilities to encourage full public use are provided. As much as 70% of the construction cost can be borne by the Federal Government in such cases. At the opposite end of the scale, where the protected shore area is privately owned and there is no public use, no Federal funds can be provided. Between these extremes, Federal participation in providing protection is proportional to public use and benefit.[3]

State and local governments or private interests must bear the remaining costs. They must also provide the necessary lands, easements, and rights-of-way necessary to construct and maintain the project and also maintain the completed works and assure continued public use of the protected shore area. Where multiple-purpose shore restoration projects include hurricane protection, the federal government will provide 70% of the costs of this portion of the project.

FOOTNOTES

[1] Nersesian, Gilbert K. "Federal Beach Erosion Control Activities on Long Island," in *Proceedings of the Seminar on Dredging and Dredge Spoil Disposal and Coast Stabilization and Protection,* Regional Marine Resources Council, Hauppauge, New York, February 15, 1973, pp. 86-87.

[2] U. S. Department of the Army Office, Chief of Engineers. *Shore Protection Program* (Washington, D.C.: 1971), pp. 5-6.

[3] *Ibid.,* p. 10.

CHAPTER 4

PROJECT DESIGN FOR WESTHAMPTON BEACH

FIRE ISLAND INLET TO MONTAUK POINT SURVEY

Almost 20 years ago, in response to mounting concern over the damaging effects of storms and erosion to Long Island beaches, the New York State Department of Public Works initiated a cooperative beach erosion control study with the Army Corps of Engineers. The state's application for the study was dated April 12, 1954, under the authority of Section 5, Chapter 535 of the New York State Laws of 1945. The Corps of Engineers had been authorized to enter into cooperative agreements for such studies by the River and Harbor Act of 1930, as amended (Section 2, Public Law 520, 71st Congress). The agreement was approved on October 14, 1954, initiating a four-year study that produced a comprehensive survey report in July 1958.[1]

Hurricane protection is closely related in federal programs to shore erosion control. Wide beaches and strong, stabilized dunes are natural protection against flood damage from hurricane and other storm surges. They protect the barrier island itself as well as lives and property on the mainland, for eroded barrier beaches and dunes expose the mainland shores to the full destructive force of the waves and tides. Great hurricanes in 1954 and 1955 caused the loss of about 200 lives and more than a billion dollars of flood damage along the East Coast. Congress then authorized the Corps of Engineers, by Public

Law 71, 84th Congress, 1st Session, 1955, to develop protective measures. For this reason the cooperative comprehensive study of the Long Island south shore included both beach erosion and hurricane protection components.

The purpose of the cooperative study was "to determine the most practicable and economic method of restoring adequate recreational and protective beaches and providing continued stability to the ocean shore from Fire Island Inlet to Montauk Point, N.Y. and to develop an adequate plan of protection against hurricane tidal flooding in the same area."[2]

On May 27, 1960, the Chief of Engineers, Department of the Army, submitted a report to the Secretary of the Army. The report incorporated the findings of the 1958 Corps of Engineers Survey and also recommendations for a beach stabilization and hurricane protection program for the south shore of Long Island from Fire Island Inlet to Montauk Point. Since the 1960 report is a succinct summary of the justification for the program and the major design features of engineering works involved, relevant numbered paragraphs are quoted here.

REPORT OF THE CHIEF OF ENGINEERS, DEPARTMENT OF THE ARMY

3. The area considered for protection consists of the beaches along the Atlantic Ocean and the areas subject to flooding from the ocean in the reach from Fire Island Inlet to Montauk Point. Its frontage, about 83 miles long, comprises about 70 percent of the total ocean frontage of Long Island. Fire Island Inlet is about 50 miles east of the Battery in New York City. Other inlets along the reach are Moriches Inlet and Shinnecock Inlet, 30 and 45 miles east of Fire Island Inlet, respectively. The normal tidal range along the area varies from 4.1 feet at Fire Island Inlet to 2.0 feet at Montauk Point.

4. The westerly 50 miles, from Fire Island Inlet to Southampton, consists of a barrier beach generally less than 2,500 feet in width topped by irregular sand dunes ranging in height up to 30 feet. In some localities the dune ridges have been partially removed or leveled and residences have been constructed on the dune line. The barrier is separated from the mainland of Long Island by interconnected bodies of tidal water known as Great South Bay, Moriches Bay, and Shinnecock Bay, which vary in width from about 500 feet to 5 miles. The bay areas are, respectively, 100, 20, and 15 square miles.

5. The 33 miles east of Southampton, which is part of the mainland of Long Island, is generally fronted by a relatively narrow beach. . . . The easterly 10 miles of this section are characterized by a series of headland bluffs rising over 50 feet above the ocean level. . . .

6. The entire area is in Suffolk County, which has experienced a 228 percent increase in population since 1930. The permanent population of the towns contiguous to the area was estimated at more than 368,000 in 1957. During the summer months the population is increased by thousands of vacationists who are attracted by the excellent recreational facilities for bathing, sport fishing, boating, and hunting. Commercial establishments on the mainland are generally of a supporting nature to the predominately recreational activities of the area. The region also is noted for its shellfish and commercial fishing industries. Truck farming, dairying, and poultry farming are the chief agricultural activities on the tributary mainland.

7. Access to the barrier beaches is provided by bridges at three places between Moriches Inlet and Shinnecock Inlet, and by passenger ferries across Great South Bay to the barrier west of Smith Point. Additional access to the barrier west of Moriches Inlet will be provided by a highway bridge now under construction at Smith Point and another being planned across Fire Island Inlet.

9. Beach and hurricane protection work undertaken by county, municipal, and private interests prior to 1946 consisted primarily of sand dunes with sand fence barriers, a seawall, and a few groins and bulkheads in isolated areas. The most extensive work was performed by Suffolk County after the 1938 hurricane when dune fill, grass planting, and installation of sand fence were accomplished over a shore length of 68 miles. Subsequent to 1946, when the State of New York initiated participation in the cost of protective works, substantial dune construction, grassing, and sand fencing has been done. In the period 1947 to 1954 the State and local governments constructed jetties and revetments at Moriches and Shinnecock Inlets at a cost of $2,692,750. In 1956 Suffolk County purchased a hydraulic dredge at a cost of $500,000 which has been used for dune restoration and channel dredging for navigation in Shinnecock Inlet and vicinity.

10. Development of the shore front in the area is predominantly summer residential and recreational in character, ranging from colonies of small bungalows and beach houses to villages of large and expensive homes and beach clubs. Buildings near the westerly end of Fire Island, about 2,500 in number, are generally of light construction, while those between Westhampton Beach and East Hampton are larger and more permanent. The most extensive undeveloped reach is that along Fire Island immediately west of Moriches Inlet, which is considered to be one of Long Island's greatest potentials for shore development. Several installations of the United States Coast Guard, the Army, and the Navy are located in the area. Public recreational developments along the shore

consist of Fire Island State Park, Hither Hills State Park, and a Suffolk County park near Smith Point, now under construction. Total investment in these parks will approximate $5,000,000. In addition, Suffolk County has recently acquired land adjacent to Moriches and Shinnecock Inlets for park purposes. The backbay shores are moderately developed for residential and recreational purposes, similarly to those along the ocean.

11. The value of lands and improvements in the entire towns bordering the ocean, exclusive of the parks, is estimated at $1.38 billion as compared to those values on the barrier plus a 0.5-mile strip along the shore east of the barrier, estimated at $118.8 million. Ownership and use of the ocean frontage is as follows:

Ownership and use	Percentage of total
Federal - public	2.6
Nonfederal - public	15.8
Private - public	2.7
Private - private	78.7

12. The shores of the area are exposed to waves of the Atlantic Ocean. For winds from the east and southeast the fetch is limited by the mainland of New Jersey. Thus, the resulting energy components produce a dominant westward littoral transport of beach material. Reversals in direction of transport materials is greater in the eastern part of the area than in the western part, resulting in less net transport in the eastern part. Intermittent surveys of the shore and offshore depths since 1834 indicate alternate erosion and accretion with a net accumulating loss of beaches. Since 1940 the net loss westward of Mecox Bay is estimated at about 300,000 cubic yards annually, resulting in recession of the beaches in certain areas ranging from a maximum of 500 feet at Tiana Beach to 70 feet at Kismet. The value of land lost by erosion is estimated at $593,000 annually.

13. Hurricane losses in the area result chiefly from hurricane tides, action of storm waves, inundation caused by hurricane-induced rain, and wind action. Records indicate that since 1635 the area was affected by 126 storms, of which 9 were unusually severe; 17, severe; 41, moderate; and 59, threats only. A recurrence of the maximum hurricane tide of record, that of September 1938 when 45 lives were lost, under 1958 conditions would cause inundation and wave damages in the area estimated at $52,600,000. The average annual ocean tidal damages in the area are estimated at $3,667,000, including $338,000 on the mainland along the inner bays.

14. Local interests desire adequate protective measures to reduce the loss of life, and beach and property damage from coastal storms. They have endorsed a general plan to provide for shore and dune restoration, artificial beach nourishment, construction of groins, construction of sand fences, and planting of dune grass. In justification of the suggested improvements, local interests indicated that: property values would be

increased; indirect losses such as loss of use of facilities, cessation of business, and interruption of traffic, would be reduced; and loss of life, beach erosion, and property damage would be prevented.

15. The District Engineer finds that the most practicable plan for protecting the area against shore erosion and hurricane damages would involve widening the beach along developed areas . . . to a minimum width of 100 feet at 14 feet above mean sea level, and raising the dunes to an elevation of 20 feet above mean sea level. . . . Grass would be planted on the dunes. . . . Fifty groins would be constructed along the shore if and when experience indicates their need. Placement of about 34,000,000 cubic yards of suitable sand would be involved initially. Maintenance of the stability of the shore would require periodic placement of about 480,000 cubic yards of suitable sand annually. The plan would require the relocation or elevation of about 80 buildings. The District Engineer believes that the best results would be accomplished if the entire plan, except the groins, were constructed as a unit. As a minimum number of separate units to be considered, he has divided the reach into five sections and has analyzed each section for costs and economic justification. On the basis of prices prevailing in June 1958, the District Engineer estimated the first cost, annual costs, benefits, and benefit-cost ratios by sections as shown in Table I. Division of first costs between Federal and non-Federal interests for beach protection is in accordance with beach erosion law, and for hurricane protection is in accordance with the cost-sharing formula in the Flood Control Act of 1958 for similar projects. Preauthorization costs in the amount of $180,000 are included in the annual costs for economic analysis but are not included in the construction cost.

16. The District Engineer concludes that the combined improvement is justified and that a Federal project is advisable. Accordingly, he recommends adoption of the project by the United States, essentially as described above, at a first cost to the United States of $18,700,000, and $15,800 annually for a period of 10 years, subject to certain requirements of local cooperation. The Division Engineer concurs.

17. The Beach Erosion Board concurs with the reporting officers in the view that the combined improvement is a technically practicable plan for the area and will provide the degree of protection contemplated. Accordingly, subject to determination by the Chief of Engineers, after review by the Board of Engineers for Rivers and Harbors, that the dual-purpose plan is suitable and economically justified, it recommends adoption of the improvement, essentially as proposed by the District Engineer, subject to the conditions that the construction of the groins be deferred pending demonstration of their necessity, and that local interests meet certain requirements of local cooperation.

18. The Board of Engineers for Rivers and Harbors concurs in general in the conclusions and recommendations of the reporting officers and with the Beach Erosion Board. It is of the opinion that the proposed

Table 1
(Costs and Benefits in $1000 Units)

Item	Fire Island Inlet to Moriches Inlet	Moriches Inlet to Shinnecock Inlet	Shinnecock Inlet to Southampton	Southampton to Beach Hampton	Beach Hampton to Montauk Point	Total
First cost						
Federal	5,902	6,325	356	5,275	801	18,700
Nonfederal	4,172	6,799	499	6,398	345	18,200
Total	10,074	13,124	855	11,673	1,146	36,900
Annual costs						
Federal	222.3	226.3	12.9	189.4	28.4	679.3
Nonfederal	433.8	399.1	32.6	354.2	28.3	1,248.0
Total	656.1	625.4	45.5	543.6	56.7	1,927.3
Benefits						
Beach loss prevention	162.1	242.1	27.6	161.1	–	592.9
Additional beach use	1,005.0	248.0	7.9	139.1	–	1,400.0
Hurricane damage prevention	1,388.5	620.9	41.2	1,075.5	176.0	3,302.1
Total	2,555.6	1,111.0	76.7	1,375.7	176.0	5,295.0
Benefit-cost ratio	3.9	1.8	1.7	2.5	3.1	2.7

dual-purpose project is economically justified, and that the improvements merit Federal participation in their cost. The Board believes that local interests should bear any additional beach protection not of an emergency nature. It concurs with the District and Division Engineers in the view that the report of the United States Fish and Wildlife Service presents insufficient data to justify establishing fish and wildlife as a project purpose. The Board concludes that the Chief of Engineers, prior to initiating construction, should make such adjustments in dredging plans as he may find feasible and economically justified to mitigate adverse effects on fish and wildlife. It states that additional costs for such purposes should be shared between Federal and non-Federal interests in the same proportions as for other construction items. Accordingly, it recommends the improvements, essentially as proposed by the District Engineer, including the special items referred to above, subject to certain conditions of local cooperation.

REACTION TO THE CHIEF OF ENGINEERS' REPORT

The Chief of Engineers' report was dated May 27, 1960. In February of that year the Chief of Engineers had sent a copy of the proposed report to the State Superintendent of Public Works for review. The report was also reviewed by the Long Island State Parks Commission and the Suffolk County Board of Supervisors. In light of the account in the next chapter of the influence of technical considerations on federal, state, and county policies regarding this project, the following comments of the DPW Superintendent should be noted:

> In general we concur with the proposal to erect sand dunes at an elevation of approximately 20 feet above sea level and plant grass thereon to prevent wind erosion, and to provide beach fill where necessary. Such a master plan for the South Shore is the only practical means of protection. . . .
>
> In our artificial beach nourishment construction we have found that where a combination of groins and fill has been incorporated, the restored beach has remained reasonably stable. For this reason we strongly suggest that the number of groins to be authorized initially should not be less than 50; and that at least part of the groin system should be constructed concurrently with fill placement.[3]

The Chief of Engineers did not modify his report to take account of these comments by the DPW Superintendent. The report to the Secretary of the Army stated in Paragraph 15: "Fifty groins would be constructed along the shore *if and when experience indicates their need*" (emphasis added), and in Paragraph 17: "The Beach Erosion Board concurs with the reporting officers [and] . . . recommends adoption of the improvement, essentially as proposed by the District Engineer, *subject to the conditions that the construction of the groins be deferred pending demonstration of their necessity. . . .*" (emphasis added).

As we shall see, the local view prevailed over the federal plan, and construction priorities were completely reversed. Instead of rebuilding the beaches and dunes first with hydraulic fill and constructing groins only if necessary to stabilize the fill, what actually happened was that groins were constructed first, and no fill was placed in the first groin field at all! The Corps of Engineers Survey of 1958 emphasized the purpose of the groins, which was to hold the beach fill in place. If they could reduce the rate of loss of beach material, the cost of artificial replenishment would also be reduced. As there had been little groin construction in the area, no data were available by which to estimate how much the rate of erosion of beach fill might be reduced. The need for groins where long stretches of coast are involved had not been definitely ascertained. Therefore, the location of groins, if they were found necessary, should be determined after the fill had been placed and there had been some experience with periodic beach nourishment.

Suggested groin design was based chiefly on the requirements for holding the beach fill in place and offering minimum interference with the littoral drift. To achieve this objective and maintain a minimum berm width of 100 feet, the groins should have been only about 400 feet long, their outer ends about 2 feet below mean low water, and spaced about 1000 feet apart. If the groins were constructed before the fill was placed, work should have started at the downdrift (Moriches Inlet) end of the beach and future construction delayed until sufficient sand had been trapped by the first groins. If the beach between the groins were to be filled immediately, however, the groin field construction could begin anywhere, for the structures would not contribute to erosion of the downdrift sections of beach.[4]

This is an important point for understanding the recent shore erosion problem at Westhampton Beach. The Corps of Engineers did not follow its own recommendation. The first phase of the project, the construction of 11 groins, was carried out in the middle of the barrier beach, within the jurisdiction of the Village of Westhampton Beach. This decision may have been taken because erosion here was more serious than on other sections of the barrier island, or it may have been politically expedient to begin the project where the investments in private resorts, beach clubs, and homes were highest. In any case, fill was not placed out to the ends of the groins. The beach berm did become wider as sand from the littoral current was trapped by the groins, but the deprivation of sand to the beach to the west of this first stage aroused demand for the second stage. Beach fill was not placed out to the ends of these four groins either, and to this omission is attributed the latest erosion problem near the western end of the barrier beach. This is the basis for the local controversy about whether six more groins should be constructed to complete the project.

FOOTNOTES

[1] U. S. Army Engineer District, New York, Corps of Engineers. *Fire Island Inlet to Montauk Point Cooperative Beach Erosion Control and Interim Hurricane Study,* Survey (July, 1958).

[2] *Ibid.,* p. 2.

[3] Letter dated May 19, 1960 in Department of Environmental Conservation files.

[4] U. S. Army Engineer District. *Fire Island Inlet to Montauk Point,* Survey, p. 37.

CHAPTER 5

TECHNICAL INFORMATION AND POLICY POSITIONS

It has already been indicated that there are scientific uncertainties about ocean beach dynamics. Some of the conflicting theories and proposals for dealing with beach erosion have been described. The Corps of Engineers considered the alternatives and settled on a design that would use hydraulic fill to rebuild the beaches and dunes at Westhampton as the first line of defense against storm and hurricane damage. Stone groins would also be constructed, but their primary function would be to stabilize the fill. Because the prevailing littoral current was from east to west, the study warned that the groin field should be started at the westerly end of the Moriches Inlet to Shinnecock Inlet reach. If the first groins were to be built elsewhere, then fill should be placed to the ends of the groins, so that they would not trap sand from the littoral current and deprive the beach to the west of nourishment.

Technical studies and project design are not always taken into account in policy decisions, however, In the case of the Westhampton Beach project, every important technical consideration was ignored in policy-making, apparently at local insistence. The rationale for disregarding these technical factors was the local desire to reduce project costs and the Corps of Engineers went along. The groin field was started in the middle of the Moriches to Shinnecock reach, at the Village of Westhampton Beach, instead of at the westerly end. The project was started with 11 groins, but no fill was placed on the beach or dunes. The unfortunate consequences might have been anticipated.

STAGE I: THE ELEVEN-GROIN FIELD

The Corps of Engineers survey was completed in 1958, and construction of the project to stabilize the shore and provide hurricane protection at Westhampton Beach was authorized by Congress in the River and Harbor Act of 1960. It was not until January 1965, however, that construction of the first 11 groins was started. Cooperative projects requiring federal, state, and local participation involve extensive and difficult negotiations. The Corps of Engineers, acting on behalf of the federal government, requires specific assurances of local cooperation, including financial contributions, provision of land and easements for construction and maintenance, and alteration of existing buildings, streets and utilities.

The federally designed beach stabilization and hurricane protection project required three-party participation by federal, state, and county governments. Actual negotiations, however, were on a two-party basis: federal-state and state-county, with the state in the middle. The Corps of Engineers communicated officially only with the state, represented successively by the Department of Public Works, the Conservation Department, and the Department of Environmental Conservation. The state had to assume full responsibility for nonfederal contributions to the project. It then had to try to induce the county to contribute its share. The county seemed to be an erratic and balky partner, however.

As time wore on and the state had still not formally presented the nonfederal commitment for the project to the Corps of Engineers, the District Engineer in September 1962 wrote to the Superintendent of the Department of Public Works urging him to provide the required assurances so that project design could get started. The storms of March-April 1962 had caused a severe breach at Westhampton Beach (Figures 11 and 12). He provided a sample form for this document, hoping that it would soon be executed and returned.[1] After almost a year of negotiations on precise wording of the assurances, which involved participation by DPW legal counsel and a resolution by the Suffolk County Legislature, the document was signed for the state by Superintendent of Public Works McMorran on

Figure 11. Westhampton Beach. Eastward view of breach caused by coastal storm of March 6-8, 1962 (photo by U.S. Corps of Engineers).

August 14, 1963. It was accepted by the District Engineer in New York City for the United States on August 20, 1963. The operational provisions of this document are as follows:

> NOW, THEREFORE, in compliance with the conditions of local cooperation above recited and provided that the Legislature of the State of New York shall appropriate the necessary funds therefore, the Superintendent of Public Works in pursuance of

Figure 12. Westhampton Beach, April 14, 1962, after emergency closure of breach. This area is now compartmented by groins 5, 6, and 7 (Corps of Engineers photo).

said Chapter 535 of the Laws of 1945, as amended, hereby assures the Secretary of the Army as follows:

(a) That the State will provide without cost to the United States all lands, easements, and rights-of-way, including borrow areas, necessary for construction of the project;

(b) That the State will accomplish without cost to the United States all alterations and relocations of buildings, streets, storm drains, utilities, and other structures made necessary by the construction;

(c) That the State will bear 48 percent of the total first cost, a sum presently estimated at $18,800,000 to consist of items listed in (a) and (b) above and a cash contribution now estimated at $16,200,000 or, if any section is undertaken separately the apportionment of the first cost will be as shown in the District Engineer's report. . . .

(d) That the State, within the scope of the authority of the Superintendent of Public Works, will hold and save the United States free from damages due to the construction work;

(e) That the State will maintain all the works and undertake periodic beach nourishment after completion in accordance with regulations prescribed by the Secretary of the Army, except that for a period of 10 years after completion of a useful nourishment unit the Federal Government would contribute dependent on conditions of public use and ownership and other changes at the time of construction, an amount for the entire project now estimated at $18,000 annually;

(f) That the State will maintain during the economic life of the project continued public ownership of the non-Federal publicly owned shores and continued availability for public use of privately owned shore equivalent to that upon which the recommended Federal participation is based;

(g) That either the State or local interests or both will adopt appropriate laws or ordinances, prior to construction, to provide for the preservation of the restored beaches and dunes and their protective vegetation;

(h) That the State will take measures to control water pollution to the extent necessary to safeguard the health of bathers.[2]

In spite of these assurances, the county was reluctant to put up its share of the cash requirements for the project. The District Engineer in New York was concerned about the news in the Long Island Press that the Suffolk County Board of Supervisors had voted down the necessary appropriation. In January 1964 he wrote to the State DPW about this action and noted that he had to appear before the Congressional Appropriations Committee to defend his part in the Fiscal Year 1965 federal budget. Congress had appropriated funds for the project for the past two years, but no progress had been made. The District Engineer would need tangible evidence of the local desire to proceed with the project if he was to be successful in getting a third appropriation.[3]

Further assurances were given, but not to implementation of the original Corps of Engineers design. The county was apparently unwilling to appropriate the full amount of money required. The theory on which the beach stabilization and hurricane project design was based was that hydraulic fill to widen the beach and build up the dune would provide the primary defense against future storm surges. Groins would be built to stabilize the fill. Local demands to reduce project cost were apparently successful, however, in persuading the Corps of Engineers to abandon the scientific rationale for the project. It agreed to building only the groins and postponing the placement of hydraulic fill.

In a Supplement to the formal Assurance of Local Cooperation, dated April 20, 1964, the Chief of Engineers recognized the need to ease the local financial burden of the project. But because of the need for early protection of the Long Island south shore, he accepted the state proposal to postpone the placement of the hydraulic fill pending determination of the amount of beach and dune stabilization achieved by the groins alone. The Chief of Engineers imposed these conditions, however:

> a. The State agrees to add beach and dune fill when and to the extent found necessary by the Chief of Engineers, but not earlier than three years after completion of the groins—unless by mutual agreement.
> b. The State will contribute the full amount of additional Federal expenditures resulting from the separation of the construction and fill phases of the project.[4]

The field of 11 groins was completed in October 1966. Only eight months later, not three years, the Corps of Engineers was already asking for the placement of the fill in accordance with the original project design. Ignoring technical considerations, uncertain as they might be, in favor of cost-cutting turned out to be unwise. On June 1, 1967, Gen. H. G. Woodbury, Jr., Director of Civil Works in the Washington, D.C. Office of the Chief of Engineers, wrote about the problem to J. Burch McMorran, Superintendent of the State Department of Public Works:

> Because of the experimental nature of the groins, this office has closely observed the functional performance of the new groin system and the behavior of the adjacent shores. In this connection, field surveys of the Westhampton area were completed in March 1967 and compared with surveys made in 1962-1964. An analysis of the volumetric changes during the period of record reveals that there has been a substantial accretion in the offshore zone between the planes of mean low water and minus 12 feet. However, despite the large net accretion above the plane of minus 12 feet there has been very little change in the beach area above mean high water and there has been no build up of the dunes to the rear of the groins. To the west of the groin field there has been heavy beach and dune erosion exposing several buildings to wave attack and has made the area extremely vulnerable to breakthroughs from the ocean. Such a breakthrough actually occurred during the storm of 28-29 April 1967 when the ocean waters flooded Dune Road west of the groins.
>
> The performance of the existing experimental groin system at Westhampton Beach demonstrates that groins alone will not fully provide the beach erosion control and probably none of the hurricane protection authorized under the existing Federal project, and that dune and beach fill is critically required at this time to supplement the completed groins. In view of the urgency of the matter as indicated by the critically eroded condition of the Westhampton area, it is considered that necessary steps should be taken by the State of New York to fulfill the conditions of local cooperation required under the Federal project and that mutual agreement to the placement of the sand fill specified in your assurance of local cooperation be implemented at this time.[5]

The Corps of Engineers' concern about the consequences of eliminating the beach fill as part of the project to build the first groin field was confirmed by subsequent events. As might be expected, the new groins began to trap sand from the shore current and deprive the beaches to the west, causing severe erosion. Properties on the beach were seriously affected, and one owner filed suit. A staff memorandum to Deputy Conservation Commissioner W. Mason Lawrence noted that Westhampton House, Inc. had filed a claim against the United States for $3 million for damage to its ocean front property on Dune Road (Figure 13):

> The claimant contends that the damage was the proximate result of negligent planning design, construction and maintenance of a certain groin or jetty to the east of the claimant's property.

Figure 13. Westhampton Beach, April 25, 1968. View of eroded shore west of groin 11. Note position of Westhampton House (Corps of Engineers photo).

> The groin named in the complaint is the westernmost in a series of 11 constructed as the first increment of the Federal-State-Local beach erosion control and hurricane protection project for Fire Island Inlet to Montauk Point.[6]

Unfortunately for the state and county, in their Assurances of Local Cooperation they agreed to "Hold and save the United States free from damages due to the construction works." Any compensation awarded by the court in this case would therefore have to be paid by the local participants in the project. As we shall see, a similar series of decisions and consequences occurred in the design and construction of the next project increment of four groins. Again the fill was eliminated at local insistence to reduce project costs, and again the result was serious erosion to the west of the new field.

STAGE II: THE FOUR-GROIN FIELD

There was a prompt response to Gen. Woodbury's letter in the form of a meeting held on June 8 and 9, 1967 of representatives of the Corps of Engineers, the state, and Suffolk County. A week later DPW Superintendent McMorran wrote to Gen. Woodbury requesting that the Corps give priority to further work at Westhampton Beach. This would not only include placement of beach and dune fill at the existing field of 11 groins, as requested by the Corps, but also the construction of an additional field of 4 groins immediately to the west (results shown in Figure 14). This was the area of critical new erosion, and the project should include the necessary fill.

Figure 14. Westhampton Beach, January 25, 1972. View of shore shown in Figure 13 after beach and dune restoration and groin construction. Relative position of Westhampton House shows extent of beach restoration (Corps of Engineers photo).

Superintendent McMorran also requested the placement of dune and beach fill to the *east* of the 11 groins. In the expectation, apparently, that matching state and local funds would be made available, he also asked that federal appropriations be continued on an annual basis until the project was completed.[7]

A meeting of federal, state, and county representatives was held in July for further discussion of this request. Suffolk County Public Works Commissioner Rudolph Kammerer did not agree completely with the state-suggested priorities. He agreed that additional groins should be constructed west of the first field because of the serious erosion there, but sand fill should be delayed. The Corps of Engineers acceded to this proposal, and tentative agreement was reached that the first contract would be for four new groins west of the completed field, and that the sand fill contract would follow shortly thereafter. This latter contract would extend the sand fill easterly into the completed 11-groin field to the extent that funds were available.[8]

The District Engineer confirmed the approval of the Chief of Engineers in Washington to the following project elements: (1) four additional groins in an area 6000 feet west of the existing field; (2) concurrent placement of beach and dune fill in this four-groin area and in the area 2000 feet immediately to the west (fill would be placed to the full design cross section as authorized in the Moriches to Shinnecock project report, except that the top of the dune would be 16 feet above mean sea level instead of 20 feet and it would have a 65-foot width at that elevation; the beach berm would be filled to the 100-foot width with the top of the berm at 14 feet above mean sea level); (3) installation of sand fences and placing beach grass on the dunes; (4) the 2000-foot-long feeder beach would be 565 feet wide at elevation 10 feet above mean sea level and located immediately to the west of the above installations.[9]

The December 1968 cost estimate for this proposal was as follows:[10]

	Total	Federal (52%)	Nonfederal (48%)
Dune and sand fill	$5,934,000	$3,085,700	$2,848,300
Four groins	2,172,000	1,129,400	1,042,600
Total	$8,106,000	$4,215,100	$3,890,900

This was too much for the local interests, so the Corps of Engineers agreed to modification of the sand fill by eliminating 1,400,000 cubic yards of feeder beach and maintaining only the authorized section for 2000 feet downdrift of the four groins.[11] Again the cost estimate was too high, and the state proposed the elimination of the 2000 feet of fill to the west of the new groins. The project would then be limited to these four groins and beach and dune fill only within the groin field. Not only would the feeder beach to the west be eliminated, but no fill would be placed at the first field of 11 groins to the east. The state "considered that this proposal represents the limit of fiscal capacity of nonfederal interests at this time."[12]

The District Engineer reluctantly agreed to these cutbacks in the project, and they were incorporated into the Second Supplement to the Assurance of Local Cooperation for the Moriches to Shinnecock project. The District Engineer noted that "the elimination of sand fill downdrift as a part of the work is considered not in the best interest of the downdrift shore. However, the critical nature of the shore requires immediate action on groin and sand placement." His revised cost estimate was as follows:[13]

	Total	Federal (52%)	Nonfederal (48%)
6000 ft of dune and sand fill	$3,262,000	$1,696,200	$1,565,800
Four groins	2,172,000	1,129,400	1,042,600
Total	$5,434,000	$2,825,600	$2,608,400

More Beach Erosion West of Stage II

The demand for economy on the part of local interests and the elimination of the feeder beach and fill in the first 11-groin field had predictable consequences. Beaches immediately to the west of the four-groin field also began to erode. This was a repetition of experience with the 11-groin field, and it aroused immediate protest from the affected property owners. Work was started on the second groin field in August 1969, and on October 7 the President of the Barrier Beach Association sent a frantic telegram to Governor Rockefeller:

> Westernmost groin at Westhampton Beach, Long Island now being constructed has produced very heavy scouring action seriously jeopardizing beach and houses to the west. Two thousand feet of fill promised by the Corps of Engineers to be placed west of the last jetty in conjunction with groins and fill urgently needed now to prevent repetition of costly destruction. Repectfully request your immediate help.[14]

Governor Rockefeller also received an irate telegram from Leonard Sacks, a resident in the affected area:

> U.S. Army Corps of Engineers construction westernmost jetty in conjunction with N.Y. State Westhampton Beach, N.Y. causing imminent destruction of my home. Demand immediate prevention measures.[15]

A week later another property owner on Dune Road, Eleanor B. Connell, wrote in a similar vein to Robert A. Cook, Director of Central Engineering in the State Conservation Department:

> As you know, there was major destruction of buildings, on both the ocean and bay side, the beach and the road, when the former set of [eleven] jetties were put in. It seems incredible that the same mistake could be made again.
>
> Since the new group of [four] jetties have been constructed, I have been taking a series of photographs. They are self-evident of the damage done to the beach west of the recent jetties. This is not an act of God but of men! Again the ocean may come through and many homes destroyed. Probably the only thing that could help protect the people and the property now is immediate fill on the beach for a mile or so beyond the last jetty. I believe 2000 feet of fill was promised by the Corps of Engineers, and it should be more as the scour has already gone beyond the area that the 2000 feet of fill would cover.
>
> It's a sad comment that the jetties were not build west to east as many experts and knowledgeable beach residents recommended from the beginning.[16]

The State Conservation Department answered all of the letters it received as well as the communications to Governor Rockefeller. DEC officials wrote in these replies that recent heavy seas had caused erosion on Westhampton Beach in the area of groin construction. They said the Department was proceeding with the Fire Island to Montauk Point project and would move ahead to future segments as rapidly as funds became available.

"Unfortunately, unprotected areas will continue to be vulnerable to erosion until a project segment can be constructed to provide protection."[17]

This letter did not satisfy Mr. Sacks and Mrs. Connell. The former wrote to W. Mason Lawrence, Deputy Conservation Commissioner:

> I tend to think that . . . you overestimate the amount of damage done by the heavy seas . . . and have underestimated the damage done directly by the groin construction.
>
> All the heavy seas did was to accentuate and dramatize how the groin can, on its easterly side, provide great benefits, and on its westerly side create great destruction. I am going to send you under separate cover an aerial view of this dramatic difference. . . .
>
> I think any objective evaluation of this situation results in the obvious conclusion that it is the groin which has taken away approximately 300 feet of beach at my house. . . .
>
> It seems to me that a condition which is directly attributable to construction in which your department is involved requires that consequences of this construction be laid at your doorstep.[18]

Mrs. Connell wrote to Robert A. Cook with the same observation about the effect of the heavy seas on the beach:

> I am aware that there were relatively heavy seas a few weeks ago, but for the Westhampton Beach *east* of the jetties it was no problem. The damage is going on *west* of the number 4 jetty (known as the scour area).[19]

The field of four groins was completed the following year, in November 1970 (see Figure 15). Since that time erosion of the beach to the west has continued. By 1973 the beach had become so narrow that a severe storm washed over the barrier island in this area. The Town of Brookhaven Neighborhood Recreation Center beach was completely destroyed and houses in the area were damaged.

Area property owners and the local municipalities still favor construction of six more groins, with appropriate beach and dune fill, to finish the Moriches Inlet to Shinnecock Inlet project. This would presumably solve the problem. Again the state and county have raised financial obstacles to proceeding with stage three of the project. At that time the state wanted the federal government to pay 70% of the total project cost for the beach erosion control portion as well as the hurricane protection portion. The County Executive did not want to spend any money at all on beach erosion control structures. The state's demands have been met by two acts of Congress. The Suffolk County Legislature has overridden two County Executive vetoes against planning the project in the County Capital Program. Perhaps all parties may move toward agreement in the near future. Meanwhile the town beach at the Neighborhood Recreation Center is being partially restored by pumping dredging spoil to it from a federal navigation improvement project in Moriches Bay.

Figure 15. Westhampton Beach, January 25, 1972. View of 15 groins constructed by Corps of Engineers (Corps of Engineers photo).

STAGE III: THE SIX-GROIN FIELD

Since the completion of the four-groin field in November 1970, property owners on Westhampton Beach west of this project have pressed for action. One of their strategies to force the issue has been to file suits for damages. In August 1973 the law firm of Skadden and Arps informed the State Environmental Conservation Department that they had filed suit against the federal government and Suffolk County on behalf of a group of property owners.[20]

In the U.S. District Court in New York, Thomas O'Grady and Dorothy Patton, individually and on behalf of all others similarly situated as members of the Barrier Beach Association, filed suit against the United States of America and Suffolk County. One of the experts called to testify was Francis R. Pagano, Chief of the Engineering Division in the New York District of the Corps of Engineers. In an affidavit dated November 11, 1973, Mr. Pagano stated that there was no pending action to build additional groins. There were no federal funds available and none were included in the proposed Fiscal Year 1975 budget. He was asked about the possibility of removing the 15 existing groins instead of building 6 new ones. He said the shoreline would tend to even out as the result of loss of beach width where the groins would be removed and accretion at the eroded downdrift beaches. Nevertheless, in the longer run the resulting narrower beaches would expose development on the barrier island to storm attack:

> Eventually some of this residential development would be destroyed or damaged. Breaching and overtopping of the barrier beach could be expected to occur during storms, accompanied by damage to the environment of Moriches Bay.[21]

Mr. Pagano estimated the cost of removing the existing groins at $7 million. This was equivalent to the 1973 estimate for constructing the six additional groins. This alternative had been suggested several times, perhaps only to indicate that the status quo of continued erosion west of the groin field was intolerable. Either the six new groins should be built or the existing field removed. Mr. Pagano warned against the latter course; and area residents and local governments would probably rise in protest against going back to pre-1965 conditions that posed constant threats to life and property from hurricanes and northeasters.

Financial Considerations

Stage III of the Moriches to Shinnecock project included the construction of six groins and the placement of dune and beach fill along 9000 feet of beach west of the existing groin field. Design of the new project had been started early in 1971, but work was suspended pending resolution of local action to change the apportionment of costs between the federal government and local interests.

In December 1972 the State Environmental Conservation Commissioner was asked by the Army District Engineer if the state intended to go ahead with the Westhampton project on a 52% federal, 48% local cost-sharing basis. Federal funds were available for the project at that time, but they would be returned to the Treasury if the state did not notify the Corps of Engineers of its intent to participate in the project.[22]

Officials in the State Department of Environmental Conservation recognized that the eroding area of Westhampton Beach was in serious condition. There was a recommendation that work on the six-groin part of the project should proceed even if a more favorable cost-sharing formula was not available. At that time, November 1973, the differences in the estimated nonfederal share of costs were as follows:[23]

	50-50 Formula	70-30 Formula
State cost	$2,450,000	$1,540,000
County cost	1,050,000	560,000
Federal cost	3,500,000	4,900,000
Total cost	$7,000,000	$7,000,000

About a year later, after intermittent negotiations, State Environmental Conservation Commissioner Diamond formally requested the Corps of Engineers to proceed with planning for the six-groin project. He wrote, "It is the State's intent to contribute the non-Federal share of the project cost. The non-Federal share is to conform to the apportionment or to any modification of the apportionment authorized by Congress."[24]

Mention of the possible modification of the cost-sharing formula was made because the state anticipated that Congress would indeed reduce the local share. In fact an act of Congress had given the Chief of Engineers discretion to apportion project costs on a federal-local 70-30 percentage basis, but he had determined not to do so for this project. The state continued to lobby for more favorable Congressional action that would mandate change in the cost-sharing formula. Senator Javits and Rep. James R. Grover, Jr., who was from Long Island, supported such an amendment to the law. This was the Water Resources Development Act, which was the new name for the old River and Harbor Act. In 1974 state efforts were successful. Section 31 of the 1974 Act (PL 93-251) amended Section 101 of the River and Harbor Act of 1960 to provide that the nonfederal share of the first cost of beach erosion control as well as hurricane protection projects would be 30 percent.

This action opened the way for further progress even though project cost estimates had risen to $10 million. In December 1973, the County Legislature authorized participation of Suffolk County in the cost of the construction of six additional groins at Westhampton Beach.[25] The county was already proceeding with necessary land acquisition, but project design would take some time. Preliminary surveys and the drafting of the environmental impact statement had not yet been started by the Corps of Engineers.[26] By May 1974, the Department of Environmental Conservation had informed the District Office of the Corps of Engineers that both the state and Suffolk County legislatures had committed funds to the project. In anticipation of federal construction funds for Fiscal Year 1975, the Corps was requested to continue preconstruction planning and design.[27]

Fish and Wildlife Service Views of the Project

In 1964 when the Moriches Inlet to Shinnecock Inlet project was in the planning stage, the Corps of Engineers contacted the federal Fish and Wildlife Service. Hundreds of thousands of cubic yards of fill would be needed to reconstruct the beaches and dunes. The source of this sand would be the bottom of Moriches and Shinnecock Bays. Disturbing the bay bottom by dredging, however, would disturb the bay ecology, especially the commercially valuable fish and shellfish resources. The Fish and Wildlife Service was therefore asked to evaluate the impact of the project on the ecology of the bays. The Fish and Wildlife Service went beyond this evaluation and considered the sportfishing potential of the groins themselves. If public access could be provided, they would produce a net average annual opportunity of 6,000 fisherman-days over the life of the project, and the annual value of this recreational activity was estimated at $12,000 in 1964.

We have seen that the Chief of Engineers, in his 1960 report to the Secretary of the Army recommending the beach stabilization and hurricane protection project, had stated that "the report of the United States Fish and Wildlife Service presents insufficient data to justify establishing fish and wildlife as a project purpose." Further public opposition was aroused in 1964 when the Fish and Wildlife Service recommendations were reported in the Long Island newspapers. One Westhampton Beach property owner's response to the proposal was a letter to Suffolk County Executive H. Lee Dennison. The letter opposed public access to the groins on the ground that easements over private property for construction and maintenance of the groins had been given to the County without cost on the basis that they would not be open to public access.[28]

Dennison replied to the effect that no public access or parking lots would be provided for fishermen. He also wrote to the State Department of Public Works: "The access areas were acquired at no cost with the stipulation . . . that there would be no public use thereof. This ends the matter as far as I am concerned."[29] The Department of Public Works, in turn, informed the Corps of Engineers that "this Department does not favor modification of the project to include the proposals of the Fish and Wildlife Service for numerous reasons, principal of which is local opposition."[30] The first groins were, therefore, closed to public access, as were likewise the publicly owned beaches below the high water line.

When the second stage of four groins was being considered, the Fish and Wildlife Service reiterated its recommendations for public access.[31] The Corps of Engineers again rejected the proposal. The Chief of the Engineering Division wrote to the Fish and Wildlife Service: "New York State advised that it did not favor modification of the project to include your recommendation . . . principally because of strong local opposition. Therefore, in view of the lack of local approval for the modification, no further action will be taken by this office in this matter."[32] This seems strange in view of documented assurances:

> That the State will maintain during the economic life of the project continued public ownership of the non-Federal publicly owned shores and continued availability for public use of privately owned shore equivalent to that upon which the recommended Federal participation is based.[33]

Furthermore, this was a condition imposed by the Corps of Engineers on the state and local participants in the project.

When the Corps of Engineers began advance planning for the six-groin project, it again consulted the Fish and Wildlife Service about the consequences of dredging fill material from Moriches Bay. The Fish and Wildlife Service, in turn, coordinated its review of the proposal with other interested state and federal agencies. For the third time it recommended public access to the groins.

The FWS report to the Corps of Engineers noted that construction of the fifteen existing groins had eliminated 3.3 miles or 21% of the 15.4 miles of beach formerly unobstructed for haul seining. The six-groin addition would remove another 1.7 miles, for a total of 32% of this reach. The impact of these losses could be made up by providing for sportfishing in the project. Rock groins are attractive to such species as striped bass, tautog, and bluefish. The Fish and Wildlife Service was particularly concerned about the almost complete lack of public access to the beach in the groin field area. It recommended that parking areas be constructed on the bay side of Dune Road at or near future groins No. 17 and No. 20. Rights-of-way for groin construction and maintenance could provide public access to the beach. Two one-acre parking lots would cost about $82,000 for land and improvements and about $500 per year for operation and maintenance.

The New York District of the Corps of Engineers forwarded this report to the State Department of Environmental Conservation.[34] It is not likely that the state will favor implementing the Fish and Wildlife Service plan. A DEC official noted that the proposal for public access to the groins is very complex. Providing parking fields would require costly land acquisitions, much more than the FWS estimate. Other facilities, such as toilets, would also have to be provided. Problems of safety caused by fishermen and others climbing out on the rock groins would probably require personnel to supervise these activities. Increased traffic on Dune Road and loss of privacy for the residents near the groins would arouse local opposition. Furthermore, construction and maintenance easements would probably not be donated by property owners, but would have to be purchased.[35]

FOOTNOTES

[1] Letter dated September 11, 1962, in Department of Environmental Conservation (DEC) files.

[2] Assurance of local cooperation for the beach erosion control and hurricane protection project for Fire Island Inlet to Montauk Point, N.Y., furnished in pursuance to the River and Harbor Act of 1960 (Public Law 86-645, 86th Congress), August 14, 1963.

[3] Letter dated January 2, 1964, in DEC files.

[4] Supplement to Assurance of Local Cooperation for the beach erosion control and hurricane protection project for Fire Island Inlet to Montauk Point, N.Y., furnished in pursuance to the River and Harbor Act of 1960 (Public Law 86-645, 86th Congress), April 20, 1964.

[5] Letter dated June 1, 1967, in DEC files.

[6] Memorandum dated June 10, 1968, in DEC files.

[7] Letter dated June 16, 1967, in DEC files.

[8] Memorandum dated July 27, 1967, by J. F. Kelley, in DEC files.

[9] Letter dated December 6, 1968, in DEC files.

[10] Letter dated December 16, 1968, in DEC files.

[11] Letter dated December 23, 1968, in DEC files.

[12] Letter dated December 24, 1968, in DEC files.

[13] Letter dated January 21, 1969, in DEC files.

[14] Telegram from Bernard Tuttleman, in DEC files.

[15] Telegram from Leonard Sacks, October 11, 1969, in DEC files.

[16] Letter from Eleanor B. Connell, October 15, 1969, in DEC files.

[17] Letter dated October 27, 1969, in DEC files.

[18] Letter dated October 29, 1969, in DEC files.

[19] Letter dated October 30, 1969, in DEC files.

[20] Memorandum, August 10, 1973, from Eldred Rich to Thomas P. Eichler in DEC files.

[21] Francis R. Pagano, affidavit dated November 11, 1973, in DEC files.

[22] Letter dated December 6, 1972, in DEC files.

[23] Memorandum, November 8, 1973, from Eldred Rich to First Deputy Commissioner Ronald W. Pederson in DEC files.

[24] Letter dated November 8, 1973, in DEC files.

[25] Resolution No. 956-1973 and Resolution No. 1007-1973. Letter January 1, 1974, from County Legislator H. Beecher Halsey, Jr., in DEC files.

[26] Memorandum, January 16, 1974, by J. F. Kelley in DEC files.

[27] Letter, May 5, 1974, from Eldred Rich to Col. Harry W. Lombard in DEC files.

[28] Letter January 14, 1965, from D. J. Schoonmaker to H. Lee Dennison in DEC files.

[29] Letter, January 15, 1965, H. Lee Dennison to Horace S. Evans in DEC files.

[30] Letter dated January 20, 1965, from H. S. Evans to Col. M. M. Miletich, District Engineer in DEC files.

[31] Fish and Wildlife Service report dated November 4, 1968, in DEC files.

[32] Letter dated November 25, 1968, in DEC files.

[33] Paragraph (f) of Assurance of Local Cooperation dated August 14, 1963, and signed by the State Superintendent of Public Works.

[34] Letter dated September 16, 1971, in DEC files.

[35] Interview with Eldred Rich, Department of Environment Conservation, August 28, 1974.

CHAPTER 6

ALTERNATIVE APPROACHES

The beach stabilization and hurricane protection project for the Moriches Inlet to Shinnecock Inlet reach of the Long Island south shore barrier islands, which has also been referred to as Westhampton Beach, was designed by the Army Corps of Engineers. The design was based on the extensive experience of the Corps and studies at its Coastal Engineering Research Center. New York State and Suffolk County agencies made no technical studies of their own. Their technical input into the Corps of Engineers' design was made in the course of local review of federal proposals.

It has already been noted that coastal geomorphology and the physics of beach dynamics are not yet exact sciences: reliable forecasts cannot be made of beach and dune behavior under varying weather, wind, and wave conditions. Nor is it possible to predict exactly what will happen to this behavior if engineering structures are placed to interfere with natural processes. Some conflicting theories of barrier beach management have been presented in Chapter 2. Other ideas presented here, in Chapter 6, were brought forward in response to the particular problems of the south shore beaches. Some of these were advocated as alternatives to the groin and fill strategy on which the Corps of Engineers design was based. Some of these alternatives are described briefly.

TEMPORARY STATE COMMISSION ON PROTECTION AND PRESERVATION OF THE ATLANTIC SHORE FRONT

Although the Moriches to Shinnecock beach stabilization project of the Corps of Engineers was authorized by Congress in the River and Harbor Act of 1960, local cooperation was not assured until 1964. The first stage of groin construction was completed only in October 1966. Meanwhile the Atlantic Ocean shore front continued to be attacked by severe storms. In response to the damage caused by the storm of March 1962, the Temporary State Commission on Protection and Preservation of the Atlantic Shore Front was established by state law. It did no original scientific or engineering studies, for it existed only a few months. It began work April 5, 1962 and completed its final report July 27, 1962. Apparently the commission relied on its consultants and earlier studies as support for its recommendations.

For the heavily developed portions of the ocean front at Coney Island and the Rockaways, the commission recommended restoring the beaches with hydraulic fill and reconstruction of the existing groins, which the report called jetties. Its proposal for the less developed sections of the Atlantic shore, as at Staten Island and Fire Island, was radically different. Its model was the Ocean Parkway that had been completed at Jones Beach. The idea was that a shore drive built on massive hydraulic fill would act like a protective dike. It would presumably replace the natural dune line and be high enough to avoid overwash by storm surges. For Fire Island, the Commission recommended public acquisition of a strip of dunes about 300 feet wide. The boulevard to be constructed on top of the dunes would have a roadway 75 feet wide. The Commission justified this strategy as follows:

> There is no excuse for spending large sums of public money to improve vacant, privately owned barrier beaches for easy minute subdivision and huge profits. Nor is it justifiable under the head of disaster to repair at public expense summer houses built too close to the ocean and badly planned boardwalks and utilities. A total of over $22,000,000 of state and local funds has been spent on beach protection since 1945. This entire expenditure has afforded only temporary relief and has been of little permanent value.[1]

This sounds something like the point of view of Suffolk County Executive John V. N. Klein about further public spending at Westhampton Beach. He prefers to avoid protecting private property at public expense and would rather let nature take its course with the beach. But Fire Island, in 1962, was vastly different from Westhampton Beach. The only access to Fire Island was by summer ferries and private boats. Developed portions of the island had relatively inexpensive summer houses without costly street and utility installations. The commission did not propose to protect such beach front properties, but rather to acquire and replace them with a costly shore-front boulevard. The boulevard would, in turn, have to be connected to the mainland by long causeways and bridges.

The commission made no specific recommendations for Westhampton Beach, but the situation there is quite different. The barrier island is accessible from the mainland by three bridges and Dune Road, which runs from one end of the island to the other. The shorefront of Westhampton Beach is almost entirely developed with very expensive homes, resorts, and private beach clubs. The section of beach where the third stage of groin construction is proposed also contains the Town of Southampton Neighborhood Recreation Center beach and parking lot. An ocean boulevard solution on top of a high artificial dune would not appear to be practical for Westhampton Beach at this time.

A wide boulevard on top of massive fill would provide protection against overwash by storm surges as long as it was separated from the active surf zone by a wide beach. Deprivation of sand for beach nourishment would eventually result in erosion of the beach to the point that the highway itself could be undermined. Unfortunately, this is exactly what happened at Gilgo Beach, a state park east of Jones Beach and on the same barrier island. Gilgo Beach has a pavillion smaller than the development at Jones Beach, but similar to it. The park is accessible by the Ocean Parkway that was supposed by the Temporary State Commission on Protection and Preservation of the Atlantic Shorefront to be a model for stabilization of the barrier islands. Nevertheless, stabilization of inlets and beach erosion control measures to the east have deprived this beach of sand.

A lack of beach nourishment along this shore has resulted in large shoreline recessions which now threaten the shore highway. Previous placement of feeder beaches under the authorized Federal project were too far eastward to permit proper functioning. In January 1969 the emergency placement of 725,000 cubic yards of beach fill at Gilgo Pavillion was completed by the Corps of Engineers to prevent a possible interdiction of the shore highway. However, about two and one-half years later, most of the fill material had been moved by littoral action.[2] [See Figures 16 and 17.]

Figure 16. Gilgo State Park, March 11, 1969. Westward view of shore in vicinity of Gilgo Pavillion after completion of emergency placement of 725,000 cubic yards of beach fill in January 1969 (U.S. Corps of Engineers photo).

Figure 17. Gilgo State Park, August 17, 1971. Westward view of shore in vicinity of Gilgo Pavillion 31 months after placement of emergency beach fill. Note significant loss of fill from the shore as shown in Figure 16 (U.S. Corps of Engineers photo).

CONSERVATION DEPARTMENT REVIEW OF PROGRAMS

During 1968, after the eleven-groin field had been completed and the four-groin field was being negotiated, the State Conservation Department proposed that the entire beach protection program be reviewed. State participation in beach protection measures had been authorized by the Beach Protection Law, Chapter 535 of Laws of 1945. This was a program of equal cost sharing with local governments for a series of relatively small projects. Some groins and sea walls were built, but most projects involved placement of sand as beach and dune fill on seriously eroded beaches.

Conservation Department review of the state-local program suggested that the state should continue its commitment to participating in shore protection measures, even though some of them were not effective.

> A number of projects constructed on Long Island have included groins or sea walls and have performed well. . . . Most of the sand fill projects in the Fire Island Inlet-Montauk Point area have been undertaken in the past to restore badly eroded areas without groins or other devices to hold the sand once it was placed. Many of these fills again have been eroded and lost in one or two years and such fill has been placed at some of these locations several times. . . . This sand fill program has resulted in a series of costly recurring projects.[3]

The state Beach Protection Law was later amended to authorize the state to participate in federal projects by sharing half of the nonfederal costs with local governments. This amendment was enacted to implement the federal Fire Island Inlet to Montauk Point beach stabilization and hurricane protection project, which required state and local cost-sharing. The Conservation Department report noted that the project "consists of the construction of a dune section with a top elevation of 20 feet above mean sea level, and a massive beach fill section. At the insistence of local interests, 50 groins were included in the project plan."[4]

Eleven groins were constructed, but extensive shore erosion occurred again immediately to the west of this field. On November 12, 1968, a storm further eroded this section of beach, destroying a restaurant and damaging several houses and a hotel. The storm essentially destroyed the existing dune:

> All along the beach in the damaged area, erosion of the dune has exposed both summer and year-round homes to potential additional damage should ever a moderate storm occur. In addition, a severe storm could cut a new inlet through the barrier beach at a number of locations. If the dunes are not restored a future hurricane might sweep across long sections of the barrier beach causing abnormal tides and waves in the inner bay, and subsequent damage to mainland developments along the northerly shore of Great South Bay.[5]

The Conservation Department memorandum asked if the federal-state-local project for the whole Fire Island Inlet to

Montauk Point reach was effective. Its answer was: "No, at least not as presently proposed." There were several reservations about the Corps of Engineers program: (1) Suffolk County felt that beach and dune fill alone was a lost cause. The massive reconstructed beach and dune would be eroded away quickly without groins. (2) Dredging the huge amount of sand needed for the project would seriously damage the ecology of the back bays and reduce the value of the shellfish harvest. (3) The estimated cost of the project had more than trebled since it was first proposed. (4) Reconstruction of the dune to a height of 20 feet above mean sea level would go beyond the requirement of state law for dune restoration. (5) Federal participation in project costs might not result in significant savings for the state and county because of the excessive scale of the Corps of Engineers design and the additional federal engineering and administration charges.

The conclusion of this preliminary 1968 review of the Corps of Engineers program was "that the protection of the beaches from Fire Island Inlet to Montauk Point can best be accomplished by a groin and fill project on a scale significantly less than that proposed by the Corps of Engineers."[6] This opinion has also been expressed by the Suffolk County Department of Public Works and engineering consultants to both the Town of Southampton and the Corps itself.[7] Nevertheless, the second stage of the project at Westhampton Beach, which consisted of the field of four groins, followed the original Corps design, except that much of the proposed beach fill was eliminated and the top of the restored dune was reduced from 20 feet to 16 feet mean sea level.

Afterward, review of the Corps of Engineers approach to beach erosion control continued. In May 1970 the State Conservation Department had arrived at a proposed alternative. It called for reducing construction and maintenance costs by building the next increment of the project in stages rather than to the full scale of the Corps of Engineers design. The reconstructed dunes would have an elevation of 17 feet above mean sea level instead of 20 feet, and the groins would be built to a lower and more economical cross-section. If future experience indicated the need for more massive groins, the second stage of construction could be carried out to raise them to the full Corps of Engineers design. The following advantages were claimed for this kind of staging of construction:

1. Reduced cost of groins and fill would permit expansion of the program and still afford a reasonably effective level of protection. More miles of shoreline could be stabilized in a shorter time with available state and local financial resources.
2. Sand from dredging of authorized inlet and navigation channels would be sufficient for a lower beach and dune fill profile without causing substantial ecological damage to the back bays. The quantities of sand required for the original Corps design could probably not be dredged from the bay bottoms without causing unacceptable ecological damage.
3. Sand fencing and vegetation could build up the dune and upper beach berm over the years, thus reducing the ultimate requirement for dredged fill.[8]

This State Conservation Department proposal was presumably intended to influence the design of the next six-groin stage of the Westhampton Beach project. As the status of this project is still uncertain, decision on the final design has perhaps not yet been made. It is likely, however, that these groins will also be built to the original Corps of Engineers cross sections.[9]

BROOKHAVEN TOWN NATURAL RESOURCES COMMITTEE

A few months after the review of the Corps of Engineers program by the Department of Conservation, the Town of Brookhaven Natural Resources Committee expressed its views about the program to Governor Rockefeller. Perhaps these comments on the project at Westhampton Beach were intended to forestall similar work on Fire Island, much of which is within the jurisdiction of the Town of Brookhaven.

In March of 1969 the President of the Committee, Michael McKeown, wrote to the Governor objecting to the spending of public funds to restore eroded private beaches at Westhampton Beach. He recognized that the state government was under pressure from local governments to salvage property values in their jurisdictions. But spending large amounts of public funds to subsidize uninsurable property was not the best form of coastal zone management. The state must assume responsibility for conserving the coastal zone, but this should be done by reducing the rate of development rather than encouraging more of it by beach stabilization efforts. Mr. McKeown wrote, "We urge you to help find a way to discourage, rather than abet the threats of real estate interests to what remains of value in New York's coastal zone."[10]

Apparently the Brookhaven Conservation Committee favored the conservation strategy described by Dolan and the Godfreys in Chapter 2. This was the strategy of letting nature take its course in the cyclical overwash and restoration of the barrier islands exemplified by the undeveloped reaches of the Outer Banks. This is why Mr. McKeown advocated halting development on the Long Island shores. He sent to Senator Jacob Javits a copy of his letter to Governor Rockefeller. The Senator, in turn, asked for comments on the letter from the State Conservation Department.

Deputy Commissioner W. Mason Lawrence replied to Senator Javits in August 1969. He repeated the familiar argument that state concern extended beyond the beaches themselves:

> The beaches and dunes along the shore of Long Island are the Island's defense from damages caused by severe storms or by hurricanes. The beaches and dunes provide protection to far more than the few houses directly adjacent to them. Should a major storm or hurricane overtop or breach the dunes, new inlets into Great South Bay or the other inner bays might be formed causing sand deposition, alteration of salinity and changes in water circulation patterns. Such events might cause major changes in the ecology of the bays. Since the bays are an extremely valuable resource of the State for the production of shellfish and clams, for recreational boating and as a part of the Atlantic waterfowl flyway, we believe that a public investment to protect the bays is justified.
>
> If during a storm, beach and dune are overtopped and destroyed in a number of places, storm waves and tides will sweep across the bay causing widespread flooding and damage to private property and public roads and installations on the mainland of the Island. For a major hurricane these damages would amount to many millions of dollars. In keeping with established government policy we believe a public investment is warranted to provide protection against such a disaster.[11]

In due course a copy of Deputy Commissioner Lawrence's letter was forwarded to Mr. McKeown. The latter replied to Senator Javits as follows:

> It is very clear from many years of beach erosion control efforts that technological capacity is simply not adequate to reliably stabilize exposed barrier beaches . . . at costs which are either politically feasible or socially desirable. . . . With regard to Mr. Lawrence's concern about the lagoons . . . comprising an "extremely valuable resource," we can only note that the State has been, and continues to be, notoriously ineffective in preserving these lagoons from duck wastes, human pollutants, indiscriminate dredging and filling of sediments, etc. It is precisely this inability of State (and local) governments to withstand local economic interests to which I alluded in my initial letter.[12]

It is curious that this letter advocating discouragement of development should have come from an organization in the town of Brookhaven. The Conservation Committee may not have represented official town policy in Brookhaven in 1969, or local policies may have changed. In any case, a newspaper article in June 1974 reported that Brookhaven Town was encouraging development on Fire Island by "handing out building variances like trading stamps. . . . Walk through one of the more popular communities on the island and you'll see freshly built rooming houses, housing additions slapped on to create rental space, homes stacked on dunes and a new two-story bar by the waterfront."[13] Another article in a different newspaper stated that

> Those arguing against development now call Brookhaven the prime offender in granting permission for future development. Brookhaven has taken the position that it cannot legally prevent construction of homes on Fire Island without violating property rights, even where zoning variances must first be granted by the town. "The courts are not going to say you can't build a house if everyone around you has a house," Supervisor Charles Barraud said.[14]

The Fire Island Association, representing about 1500 residents, filed suit against the local governments, including Brookhaven, that control building on the island, and against the National Seashore administered by the Department of the Interior. The suit was filed in federal court in 1972 to limit new construction on Fire Island to single-family homes on regulation-sized plots until the Fire Island National Seashore publishes a master plan and an environmental impact statement. The Fire Island Association contended that the 1964 federal law establishing the National Seashore was being violated by the defendant governments, which were allowing unrestricted development on Fire Island. The suit was rejected by the U.S. Court of Appeals in New York City in May 1974.

Although this case concerns Fire Island rather than its neighboring barrier island to the east, Westhampton Beach, it illustrates the problem of limiting urbanization in the interest of conserving environmental and ecological values. What the federal courts have said is that, as far as the Long Island barrier beaches are concerned, limiting population growth and urban development in order to let nature take its course and thus avoid engineering works to stabilize the shoreline is not a viable policy.

PROPOSALS FOR PERMEABLE GROINS

Instead of objecting to engineering works altogether, some critics of the Corps of Engineers' approach to beach erosion control propose alternative groin systems. Donald M. Burmister was an engineering consultant to the Town of Hempstead on Long Island when he prepared a 1969 report on the Fire Island Inlet to Montauk Point program of the Corps. Copies of the report were sent to John V. N. Klein, then Chairman of the Board of Supervisors of Suffolk County, and Governor Rockefeller. The governor sent a copy to the State Department of Conservation.

Dr. Burmister contended, as others have, that beach erosion control is still not an advanced field of engineering, and that the mechanics of transportation of sand and water on the beaches and in the littoral current are not yet fully understood. Furthermore, conditions of sand grain size and beach berm and foreshore slope are not the same from one end of the barrier island system to the other. This requires careful selection of fill materials and design of beach fill cross sections if fill is not to be washed away in a relatively short time. In view of these uncertainties, Dr. Burmister thought that the Corps of Engineers' approach was naive and unrealistic and that the program "should be scuttled, once for all." [15] The main points of his analysis of the problem and suggestions for alternative groin design are summarized in the following paragraphs.

Dr. Burmister thought that the objective of stabilizing the entire south shore from Fire Island Inlet to Montauk Point was wrong. The 30 miles of eroding bluffs from Southampton to Montauk Point is "the essential and only source of beach sand for all beaches westward to Rockaway Point. . . . This basic fact rules out definitely and conclusively any protection of this vital beach sand source."[16] Instead of trying to stabilize this easternmost reach of the south shore, Dr. Burmister's proposal would be to compensate property owners and relocate them away from the area. Eminent domain and disaster relief rather than erosion control are his suggested strategies.

There was comparatively little development of beaches and resorts on Long Island before World War I. Prevailing natural forces maintained the south shore beaches with abundant supplies of sand from eroding headlands attacked by winter storms (see Figure 18). However, when updrift beaches began to be developed, beach stabilization works were installed. These deprived the downdrift beaches of sand, and they began to erode. Protective groins were then installed on these downdrift beaches, and the cycle of beach erosion spread westward.

The present trend of groin construction is to build long, impermeable structures to trap and impound the sand stirred up in the surf zone or by intercepting the littoral current. This is why erosion takes place on the westerly beaches. It is possible, however, to use the constructive power of the sea for moving and depositing sand. If structures are built

> so that the waves can harmlessly break and roll up a beach and expend their energy before reaching the high water mark, it is possible in many cases not only to stop erosion, but also to shift the cycle to accretion of sand on the beaches, provided that the control works do not interfere with or upset the balance of sand movements.[17]

Such structures have already been used successfully. They are permeable groins, which were developed by the Army Corps

Figure 18. Crane Neck Point, Long Island. This headland in the town of Brookhaven is undergoing the same kind of erosion as the Montauk bluffs, which furnish sand nourishment for the south shore barrier beaches (U.S. Corps of Engineers photo).

of Engineers for control of Mississippi River floods during the 1930s. Permeable jetties caused the river to build out its banks to the ends of the jetties, with no interference with river flow. Permeable wood groins were also used on beaches in the Florida Keys. They were installed by the homeowners themselves by jetting the timbers into the sand. Dr. Burmister's report contained photographs showing the successful accretion of sand by these "picket fence groins." The advantages of the permeable groin are as follows:

> 1. It does not stop the flow of oblique sand-bearing waves running up a beach, but only slows them sufficiently to cause deposition of sand.
> 2. It does not deflect the sand-carrying littoral current out to sea, where it loses its sand-carrying capacity.
> 3. As a consequence, these actions cause the beach to build up uniformly (not saw-toothed).
> 4. The permeable groins allow the free, uninterrupted movement of the drift of sand by oblique waves to downdrift beaches, thus ensuring sufficient sand for maintaining stable beaches to the west.
> 5. When the sand beach has built up to the top of the low permeable groin, it goes out of action, leaving a clear uninterrupted stretch of stabilized beaches.
> 6. If conditions change . . . so that the groin becomes uncovered, it can again go effectively into action to cause accretion of sand, stabilize the beach, and bury itself again.[18]

The idea of permeable groins was advanced later on in the State Conservation Department itself, although in a different form than that proposed by Dr. Burmister. In May 1970 a Technical Staff Memorandum was prepared by Peter Maier that described theories about the movement of sand grains in the surf both parallel and perpendicular to the beach and about the relation between sand grain size, the equilibrium profile of the beach, and erosion of fill. In light of these theories Maier noted that

> whether the groin should be impermeable or permeable is a highly debated question. It is felt that the impermeable groin has shown to be more effective, provided it is well designed.
>
> At the present time the sand movement in the breaker zone is extensively under study in laboratories as well as in the field. . . . It is expected that the results will give a clearer view on the effectiveness of either the permeable or impermeable groin.[19]

Nevertheless, Maier did conclude that from information now available, the hypothesis may be advanced that groins should have two layers. The lower one, about half the depth of the entire groin, should be impermeable, while the upper layer should be permeable. Under calm conditions, the impermeable lower layer would collect sand from the littoral drift. Under stormy conditions, the lower layer would still collect sand,

and the upper permeable layer would create turbulence in the shore current, decreasing its velocity. The permeable groin structure would also help to dissipate wave energy. When there were severe storms, the groin would have little effect either on the waves or shore current. This kind of two-layer groin could have a massive rock body for the impermeable part and rows of piles for the permeable part. Peter Maier was much less positive about the advantages of permeable groins than Dr. Burmister, and was not sure they could be justified economically.

NASSAU-SUFFOLK REGIONAL PLANNING BOARD

The Nassau-Suffolk Regional Planning Board has established the Regional Marine Resources Council to provide it with information on the extremely valuable marine resources of Long Island. With financial assistance from the New York Sea Grant program, the Council has conducted extensive studies of the marine environment to inventory its resources, identify problems of resource management and conservation, and develop policy and planning guidelines. A report on the shoreline erosion problem recommends the following policy guidelines for the south shore:

1. Encourage programs to preserve and enhance the natural capability of the barrier islands to protect the environment of the back bays from sudden changes caused by storm breaching. Place primary emphasis here on dune stabilization and beach nourishment techniques.
2. Encourage projects to stabilize existing inlets at approximately their current dimensions and location. Assure that stabilization techniques chosen include provision for adequate sand on downdrift beaches. Avoid substantial changes in these inlet characteristics unless explicitly justified by an analysis of the changes in the back bays.
3. Encourage land use management measures to control the use and development of the barrier beaches in a way that reflects the public values involved.[20]

For the particular problem of beach erosion, the report suggests these potential solutions:

1. Engineering techniques: particularly beach nourishment, dune improvement and stabilization with vegetation, inlet stabilization and sand by-passing, and some groins and local armament. . . .
2. Management techniques: particularly land-use planning maps, regulatory controls such as zoning and building codes, flood-plain management approaches such as delineation of hazardous areas, and storm warning services.[21]

In 1973 the Nassau-Suffolk Regional Planning Board published its *Guidelines for Long Island Coastal Zone Management.* The guidelines were intended for use by county and municipal governments in arriving at public policy decisions. The Regional Planning Board did not mean simply to offer these guidelines as suggestions, but rather to advocate their adoption:

> The guidelines will be forwarded to the Nassau and Suffolk County Executives and the appropriate legislative bodies . . . for consideration. An effort will be launched to convince the 107 cities, towns and villages of the Nassau-Suffolk region, which retain basic land-use control powers, to endorse and use the guidelines as a basis for decision-making.[22]

With regard to the problems of coast stabilization and protection, the guidelines advocated reliance primarily on land use and development controls, such as flood plain zoning, prohibition of construction on primary dune lines, and land-use zoning. For dealing with problems of beach erosion particularly, the Regional Planning Board favored the following policies:

> As a general rule, discourage expenditure of public monies for the design and construction of shore protection work and beach nourishment on private lands unless public access to those lands is provided.
>
> Emphasize dune stabilization and beach nourishment techniques as the primary means of minimizing storm breaching of the south shore barrier islands and thus protecting the environments of the south shore bays from sudden short-term changes.
>
> Stabilize existing south shore inlets . . . at approximately their current dimensions and locations. . . .
>
> Foster implementation of Federal projects for sand bypassing systems at Shinnecock, Moriches, and Fire Island Inlets.
>
> Prohibit the construction of groins and other shore protection devices either by government or private persons unless it can be demonstrated that such structures will not adversely affect adjacent property.[23]

These guidelines were based on information about the scientific state of the art for beach erosion control. The Regional Planning Board has not been asked to study the Westhampton Beach problem specifically or to make policy and technical recommendations for dealing with it. The Board has taken no official position about whether or not the six additional groins in question should be built, nor has it suggested erosion control action alternatives at Westhampton Beach.

One of the senior staff members of the Marine Resources Council, DeWitt Davies, grants that more scientific knowledge is needed before the effects of stone groins on particular beaches can be reliably predicted. Nevertheless, he feels that action would be warranted now. Past experience indicates that a hurricane or severe storm could again breach the barrier island at the Town Neighborhood Recreation Center at Westhampton Beach and cause serious problems for the island itself, for Moriches Bay behind it, and for the bay shore. Davies has noted the alternatives half-seriously suggested by other observers: either complete the project by building the last 6 groins, or remove all 15 existing groins and rely on artificial sand nourishment to maintain the beaches and dunes.[24]

FOOTNOTES

[1] Temporary State Commission on Protection and Preservation of the Atlantic Shorefront. *Final Report* (Albany, New York: 1962), p. 18.

[2] Nersesian, Gilbert K. "Federal Beach Erosion Control Activities on Long Island," in *Proceedings of the Seminar on Dredging and Dredge Spoil Disposal and Coast Stabilization and Protection*, Regional Marine Resources Council, Hauppauge, New York (February 15, 1973), p. 88.

[3] Memorandum, December 9, 1968, from R. A. Cook, Director, Office of Central Engineering to W. Mason Lawrence, Deputy Commissioner.

[4] *Ibid.*

[5] *Ibid.*

[6] *Ibid.*

[7] See account of interview with Robert Grover of Greenman, Pederson and Associates, p. 104.

[8] Letter, May 18, 1970, from R. A. Cook, by Eldred Rich, to G. H. Von Gunten, N.Y. District, Corps of Engineers and draft memorandum of the same date by Eldred Rich in DEC files.

[9] Interview with Eldred Rich, State Department of Environmental Conservation, August 28, 1974.

[10] Letter dated March 17, 1969, in DEC files.

[11] Letter dated August 18, 1969, in DEC files.

[12] Letter dated September 11, 1969, in DEC files.

[13] McKenna, Ken. "Imperiled Fire Island: Is U.S. Aid Helping?" *New York Daily News* (June 6, 1974), p. 72.

[14] Fresco, Robert. "Fire Island Case Appeal Set," *Newsday* (June 6, 1974).

[15] Burmister, Donald M. "Report of Beach Erosion Control for the Long Island Barrier Beaches from Montauk Point Westward to Rockaway Point," Report SM-356-1968 (September 16, 1968), p. 4.

[16] *Ibid.*, p. 6.

[17] *Ibid.*, p. 16.

[18] *Ibid.*, pp. 17-18.

[19] Maier, Peter. Technical Staff Memorandum. "General Considerations for Construction in the Surf Zone" (May 18, 1970), p. 6.

[20] Bartholomew, F. L. and W. V. McGuiness, Jr. *Coast Stabilization and Protection on Long Island*, Regional Marine Resources Council, Nassau-Suffolk Regional Planning Board (February 1972), pp. 42-43.

[21] *Ibid.*, pp. 41-42.

[22] Nassau-Suffolk Regional Planning Board. *Guidelines for Long Island Coastal Zone Management* (1973), pp. 2-3.

[23] *Ibid.*, p. 5.

[24] Interview with DeWitt Davies, June 18, 1974.

CHAPTER 7

SUFFOLK COUNTY POLICIES

Of the local governments involved in the problem of whether or not to complete the groin field on Westhampton Beach, only Suffolk County has a decision-making role. This is because the county must share with the state and federal governments the first cost of constructing the project, and must commit itself to maintaining the groins, beaches, and dunes thereafter. Suffolk County participation in the entire Moriches to Shinnecock beach stabilization and hurricane protection project has been sporadic. When the project was first conceived, the County Executive and Legislature heartily endorsed it, perhaps because hurricane and storm damage were fresh in their memories. In later years attitudes of county officials were variable, sometimes approving and sometimes disapproving further participation.

THE COUNTY LEGISLATURE

An account of these variations may be drawn from the Proceedings of the Board of Supervisors and its successor, the County Legislature, the Capital Program, and the Capital Budget. Before the first stage of the project could begin, the county had to give the state and federal governments assurance that it would cooperate in providing financial and other inputs. The project was given official status in the Federal River and Harbor Act of 1960, but not until 1963 did intergovernmental negotiations reach the point where Suffolk County would commit itself. In Resolution No. 365-1963, the Board of Supervisors offered the following commitments to the entire Moriches to Shinnecock project, which were required by the federal authorities:

1. Provide all necessary lands, easements, rights-of-way, and borrow areas.
2. Do all necessary alterations and relocations of buildings, streets, storm drains, utilities, and other structures.
3. Bear 24% of the total first cost, including the above contributions, for a total county contribution estimated at $18,800,000.
4. Hold New York State and the United States free from damages caused by construction.
5. Maintain all works and undertake beach nourishment as prescribed by the Secretary of the Army.
6. Retain its own shores in public ownership and maintain continued availability of the privately owned shore for public use.
7. Adopt appropriate laws prior to construction, in conjunction with local interests, to preserve the restored beaches and dunes and their protective vegetation.
8. Take measures to control water pollution to safeguard the health of bathers.

The following year the Board of Supervisors adopted a series of implementing resolutions. Resolution No. 75-1964 approved federal construction plans for 13 groins and requested federal and state agencies to proceed with construction of 11 of them. Resolution 210-1964 authorized the preparation of maps for the project and transferred $25,000 from the General Contingency Fund for this purpose. Resolution No. 210-1964 specifically agreed to share in financing the construction of the first 11 groins. The state would advance the entire nonfederal share, and the county would later reimburse the state for half of this amount. At that time the total project was estimated to cost $4,512,000, with the county share to be $1,128,000.

Resolution No. 302-1964 approved the maps prepared by the County Department of Public Works and called for a public hearing on the acquisition of properties necessary for construction of the project. These properties were not to be acquired as public land but only as easements for construction and future maintenance of the groins. At the hearing on June

13, 1964, Wilson F. Reynolds, then Mayor of the Village of Westhampton Beach, objected to opening these easements to public access because this would create problems for the residents of the area. Town of Southampton Supervisor Stephen F. Meschutt said that unrestricted rights-of-way would create traffic problems, and John Curley, attorney for Westhampton Beach property owners, also objected to opening public access to the beach for bathing and fishing.

The County Board of Supervisors agreed with these points of view expressed at the public hearing. This was at least partly motivated by the suggestion that the Village of Westhampton Beach and the owners of property along the beach would donate the required easements to the county if they were closed to general public access. In its Resolution No. 364-1964, the Board of Supervisors declared that the 25-foot-wide easements from Dune Road to the stone groins "be limited to use solely and only by governmental agencies and their duly authorized representatives, employees and contractors for the sole purpose of inspecting, maintaining and repairing said stone groins, with the control over said parcels to remain in the owners thereof."

This apparently contradicts the Board's Resolution No. 365 of the previous year in which it committed itself to maintaining public nonfederally owned shores in public ownership and continued availability for public use of the privately owned shore. If there is no public access to the beach, there is no way for the public to use and enjoy it. Long stretches of the beach are blocked to public access by continuous private ownership by homeowners, beach clubs, and resorts. Such occasional public access points as there are from Dune Road are restricted by permit to residents of the Town of Southampton. This limited definition of "public" and closing of the county easements to public use were not objected to by the federal authorities in spite of their own conditions for county participation in the project. At a subsequent meeting of the Board of Supervisors, Supervisor Griffin noted that before an erosion control project in Perth Amboy was approved by the federal government, it required that access to the groins and shore be open to the public. He wondered why this requirement was waived in the Westhampton Beach project. The proceedings contain no answer to this question or further comment.

As the proposed groins would occupy land under water, the county had to obtain the rights to this land from the state. Resolution No. 621-1964 authorized the county to submit an application to the State Commissioner of General Services for the grant of title to state lands under water, which were described on identified surveys. The county thus fulfilled its commitments to provide the necessary lands and financing to support the first stage of the project, and construction of the 11 groins began.

The continued urgency of beach stabilization and hurricane protection measures was reinforced by the effects of severe storms during January 1966. In Resolution No. 39-1966, the Board of Supervisors noted that severe damage had been suffered on the north and south shores of the county, and it authorized its Committee on Public Works to find out what federal and state financial aid might be available and to apply for it. In Resolution No. 626-1966, the board authorized the sale of $200,000 of general obligation bonds to finance emergency measures to deal with the storm damage. During the following year, 1967, the county continued its requests for state aid. It asked the Superintendent of Public Works to prepare plans and specifications for emergency dune restoration at Westhampton Beach and pledged itself to pay half the cost.[1] The county also asked the governor for financial help and asked him to apply for further aid from the federal Office of Emergency Planning and the Corps of Engineers.[2]

When the Corps of Engineers was prepared to go ahead with the beach stabilization and hurricane protection project in 1968 by building four additional groins to the west of the first field and using hydraulic fill from Moriches Inlet for beach fill and dune construction, the county renewed its pledge of participation. It would again have to bear half of the nonfederal costs, for an estimated county obligation of $670,000.[3]

Property maps were again prepared and approved and a public hearing was held on August 12, 1968. The attorney representing the Barrier Beach Association and the Dune Road Property Owners Association said that these organizations favored completion of the entire Moriches to Shinnecock project, not just a piecemeal extension. The property owners were also willing to donate easements for project construction and maintenance. The mayor of Westhampton Beach Village urged the county to start the work as quickly as possible. Use of the easements was again limited to government personnel and contractors for inspecting and maintaining the groins and excluding public use.[4]

Suffolk County financed its share of the first two stages of the beach stabilization and hurricane protection project by selling serial bonds. The bond issue for the first stage of 11 groins totaled $852,000, but only $841,540 was needed to reimburse New York State for its advance of the total nonfederal share. Nevertheless, by 1984 the county will have paid out $1,201,419 for principal and interest. The second project stage of four groins required a bond issue of $780,000 and a General Fund Transfer of $42,050. By the time the bonds are paid off, principal and interest will have amounted to $1,392,612. The first 15 groins and associated beach fill and dune construction will therefore ultimately cost the county about $2,636,000 for its roughly 25% share of the total project cost.[5]

The Controversial Six-Groin Third Stage

By the time consideration had to be given to the third stage of the project, the Board of Supervisors had been succeeded by the County Legislature. County Executive H. Lee Dennison had been succeeded by newly elected John V. N. Klein, and a new capital program and capital budget procedure had been instituted.

In his proposal to the County Legislature for the 1971-1973 Capital Program, County Executive Dennison included items of $700,000 for 1971 and $750,000 for 1972 for beach erosion and hurricane protection projects. He supported this request with the following statement:

> The County Board of Supervisors committed the County to participate in the erosion control project at Westhampton Beach sponsored by the Army Corps of Engineers at 25% of the total cost. Currently 11 groins have been completed and the County Legislature agreed to provide its share of the construction of four additional groins in 1970.

> The analysis of the results of this project will be incomplete unless the groin field is extended to Moriches Inlet. We are recommending in the 1971 and 1972 Capital Programs costs for the County share of the remaining groins. This program will not only complete the project and allow for thorough examination of its results, but also protect the recently acquired county land on the barrier beach.[6]

The County Legislature, however, deleted these items from the final approved Capital Program.[7]

The following year in his submission of the Capital Program for 1972-1973, County Executive Dennison apparently changed his mind and agreed with the legislature that county funds should no longer be spent on beach erosion projects: "Suffolk County's 800 miles of Sound, Bay, and Ocean shore front is perhaps the County's greatest natural resource. . . . These same beaches, however, are also one of the County's greatest problems. On several occasions, particularly in 1938 and 1955, miles of beachfront and homes on the shore have been destroyed by hurricanes."[8]

Nevertheless, county policy opposed further spending on erosion control. Federal and state funds would be welcome, but the county share would have to be borne by the residents of the area who would be benefitted by the project:

> Erosion control has become a significant financial burden for the County's taxpayers. Since 1960 the County has expended or authorized for expenditure $4,080,050 for erosion control projects assisted by additional funds from the State and Federal Governments. In 1969, the Board of Supervisors determined that the county would no longer appropriate funds from its general tax levy to cover erosion control work other than to complete projects previously authorized. This decision was confirmed by the current County Legislature and County Executive.
>
> The 1972-1974 Capital Program includes no recommendations for new projects in the area of erosion control which utilize county funds. I have included funds to complete the previously authorized federal groin construction project at Westhampton Beach.[9]

THE COUNTY EXECUTIVE

The third stage of the Moriches to Shinnecock project which was for the six final groins and beach and dune restoration, was included in County Executive Dennison's last Capital Program, and approved by the County Legislature as a "previously authorized project." Through 1973, this program proposed $32,050 from a General Fund transfer and $470,000 from the sale of serial bonds, a total of $502,050 for the county share of stage three. This amount was no longer 25% of total project cost. By this time the participation formula had changed to 70% federal, 21% state, and only 9% county. The amount in the 1972-1973 Capital program was, therefore, 9% of the estimated $5,578,420 total project cost.

There was a radical change in County Executive policy when John V. N. Klein took office. He was adamantly opposed to further county spending on beach erosion control projects, while the County Legislature still favored completion of the Westhampton Beach project. This led to a seesaw battle between Executive and Legislature that had not been resolved at the time this report was written.

In his message to the County Legislature submitting his first Capital Program, for 1973-1975, Klein had this to say:

> It will undoubtedly come as no surprise to anyone that I have recommended the elimination of the jetty construction program for erosion control at Westhampton Beach, in particular, and throughout the County in general. Notwithstanding all the arguments which have been mounted in respect to the effectiveness of these devices, several factors have become patently clear to me based on eight years of observation in county government. Among those factors is the clear indication that while the jetties may provide some source of relief for the areas which they immediately serve, they create immense problems in the direction of the flow of the littoral drift, which on the South Shore is east to west. With respect to the Westhampton Beach and East Hampton projects, the County is now faced with millions of dollars in claims for damage created by the construction for which the county is "holding the bag" because as a condition to these programs, the County must agree to hold the Federal and State Governments harmless from such claims. A recent study by Southampton College has generated an interesting concept which is that the opening of the Shinnecock and Moriches

> Inlets in the late 1930s and the so-called "stabilization" of those inlets has provided an opportunity for the ingestion of sand by those two inlets which accumulates on the interior waterways immediately inside the inlets. The effect is, therefore, to starve the beaches to the west of this sand, which would ordinarly find its way along the beach area through the littoral drift. Observation and participation in conferences involving State and Federal officials over the years have convinced me that no one, irrespective of their position in government and professional persuasion, really knows the causal factors which have accelerated the rate of erosion on both the North and South Shores of Long Island. I am equally convinced that the lack of expertise on this subject is matched by a similar lack of professional capability to deal with the problem. In a time when the residents of the Southwest Sewer District are required to shoulder the full burden of the local cost of the sewer program without participation of the County general taxpayers, it is totally unfair to expect the County to pay any money whatever to erosion control measures designed to protect private property.[10]

On the basis of this position, John Klein eliminated the items for the Westhampton Beach project that had been included in Lee Dennison's previous Capital Program. The County Legislature, however, disagreed with Klein and voted to adopt the Capital Program with provision for a $20,410 General Fund transfer and $310,000 in serial bonds for the project.[11]

During 1973 a similar series of Executive disapprovals and Legislative overriding approvals occurred. In his Capital Program submission for 1973-1975, John Klein again omitted an item for the Westhampton Beach project. The Legislature, however, in Resolution No. 489-1973 stated:

> WHEREAS, this County Legislature desires to adopt a project not included in the County Executive's recommended Capital Program, now therefore, be it
> RESOLVED, that the 1973 Capital Program be and it is hereby amended as follows:

Project No.	Project Title	Revised 1973 Capital Program
5317	Construction of Additional Groins at Westhampton Beach	$20,410 General Fund Transfer
		$320,000 Serial Bonds.

This Resolution was disapproved by John Klein in a long veto message. He noted that no funds were then available for the 91% federal and state shares for the project and that the estimated cost of $600,000 per jetty was too low. He also said that his objections to the continuation of the project could now be supported by two recent technical studies. One was the National Park Service "Dune Stabilization Study," Natural Science Report No. 6. His quotations from this report included these statements: "The results of man's past efforts to stabilize the primary interfaces of beaches and coasts have been mostly negative, resulting in more serious management problems than existed with the natural state. . . . Management strategies based on the concept of stable features are in conflict with nature."

Klein's second authority was "Erosion of the North Shore of Long Island," Technical Report No. 18 issued by the Marine Science Research Center at Stony Brook. It stated that: "Man's interference with the shoreline . . . has caused the configuration of the shoreline to change. . . . In all cases this change has not been to the benefit of man. . . . Engineering structures, such as groins, should not be constructed by government or private interests without sufficient knowledge of the processes affecting the area to insure that such structures will not increase erosion rates of adjacent property."

The Legislature was not convinced. It adopted Resolution No. 554-1973, which

> RESOLVED, that Resolution No. 489-1973 duly adopted by this Legislature on July 19, 1973, which said resolution was disapproved by the County Executive on July 30, 1973, be and hereby is adopted effective immediately notwithstanding said disapproval of the County Executive.

This was in August 1973. In December, the Legislature passed another Resolution, No. 956-1973, authorizing participation of Suffolk County in the cost of the construction of six additional groins on the shorefront of the Atlantic Ocean in the towns of Southampton and Brookhaven. This resolution was based on the premises that the 21-groin project originally designed by the Corps of Engineers should be completed, and that Section 101 of the "Environmental Bill of Rights" of the Suffolk

County Charter stated the basic policy of the county "to take all reasonable steps to conserve and protect its wetlands and shorelines."

Again there was a veto message from County Executive Klein. He disagreed that the beach erosion control project fell within the basic policy of the Environmental Bill of Rights. He said that "such a conclusion is not only contrary to the spirit and intent of the Environmental Bill of Rights, but is also contrary to the findings and recommendations of the Council on Environmental Quality and the Nassau-Suffolk Marine Resources [Council] who have reviewed the proposal within the framework of Section 101 of the Suffolk County Charter."

This second veto message stated that the proposal for the last stage of the Westhampton Beach stabilization project had been reviewed by the Regional Marine Resources Council, which opposed it. This seems to be an error, for the Council had not been asked to review this particular project.[12] It therefore appears that County Executive Klein was drawing unwarranted conclusions about this particular project from generalizations found in Council reports.

The Council report he cited simply recommends that groins should not be built unless it is demonstrated "that such structures will not increase erosion rates of adjacent property." Presumably if such a demonstration can be made, the groins would be justified. In the Westhampton Beach case, the last groins would be built at the western end of the Moriches to Shinnecock barrier beach, where there would be no occasion for further beach erosion to the west.

If nature were to be allowed to take its course as Mr. Klein proposes, Moriches and Shinnecock Inlets would be allowed to silt up rather than be stabilized since their opening by storms decades ago. It has been decided as a matter of county policy that these inlets should be kept open; hence, Shinnecock Inlet will continue to remove sand from the littoral drift. The 15 existing groins also trap sand from the shore current. Because they deprive Westhampton Beach of considerable sand nourishment, the combined effect of the inlet and existing groins would seem to pose a continuing threat to the beach west of the existing groin field.

In any case, Mr. Klein's arguments were not persuasive for the Suffolk County Legislature. It passed this resolution to override the veto:

> Resolution No. 1007-1973 adopting Resolution No. 956-1973, authorizing participation of Suffolk County in cost of construction of six additional groins on the shorefront of the Atlantic Ocean in the Towns of Southampton and Brookhaven dated December 27, 1973.
>
> Resolution No. 956 is adopted effective immediately notwithstanding disapproval by the County Executive.

THE COUNCIL ON ENVIRONMENTAL QUALITY

The Council on Environmental Quality was established by the Suffolk County Charter. Its nine voting members include eight appointed by the County Executive and the Chairman of the Park Committee of the County Legislature. The County Legislature must approve the appointments. *Ex officio* members are the Director of Planning, the Commissioner of Public Works, the Commissioner of Environmental Control, the Commissioner of Buildings and Grounds, and the Commissioner of Parks. Because of its membership, the Council is in an excellent position to coordinate its investigations of environmental issues and problems with the major county agencies.

One of the functions of the council is to review and appraise any project or activity that may affect the quality of the environment of Suffolk County, and to report its findings to the County Executive and County Legislature. The council is also responsible for setting up guidelines for environmental impact analysis by county agencies. Its reviews of these statements are forwarded to the County Executive and the County Legislature. The council is financed from the budget for the Office of the County Executive, and it is administratively related to the Deputy County Executive for Environmental Matters.

County Executive Klein had been opposed to beach stabilization by groins or other engineering works long before his election to that office and before the Council on Environmental Quality was established. Eventually, H. Lee Dennison, former County Executive, was appointed to the council, and brought to it his own position favoring the completion of the groin field on Westhampton Beach. Before that time, however, the council had looked into the beach erosion problem and arrived at a position similar to that of Klein. At its meeting of April 18, 1973, it adopted this Resolution on Development and Stabilization of Exposed Shorelines:

> WHEREAS, recent attention has been focused on the seriousness of coastal erosion on Long Island; and
>
> there is a continuous long-term rise of water level against the land, amounting to several millimeters per year; and
>
> the continental margins are continuously eroding at rates that are appreciable in decades; and the erosion means that barrier beaches must move landward; and attempts to stabilize them are assured of failure in future years; and
>
> there is convincing contemporary research showing that the stabilization of dunes prevents their landward migration and results in breaching of the barrier beach; and
>
> contemporary efforts at stabilization through beach nourishment, construction of jetties and stabilization of dunes have yielded questionable benefits,
>
> NOW, THEREFORE BE IT RESOLVED that the Council on Environmental Quality urges the County Executive and the County Legislature to adopt a policy to seek every means of:
>
> - encouraging restriction of further development of exposed shoreline, and
> - discouraging attempts at stabilization of intrinsically unstable areas.

The council does not have the staff or financial resources to carry out its own empirical studies of beach erosion problems. It appears to have relied on its evaluation of recent literature and the professional experience of its own members in arriving at the conclusions expressed in this memorandum. This literature, as summarized by a staff member of the council, includes only studies similar to those of Dolan and the Godfreys cited in Chapter 2 that favor allowing natural processes to shape the physical configuration of barrier islands. These authors would not interfere with periodic washovers or the opening and closing of inlets by seasonal weather cycles and severe storms. They conclude that barrier islands are inherently dynamic systems, and that attempts to stabilize them by engineering works are doomed to failure. Furthermore, they believe that empirical evidence shows that barrier islands are strengthened rather than weakened by washovers and inlet changes, and that shellfish production in the back bays is increased by periodic changes in salinity and temperature of the water there.[13]

By the May 23, 1973 meeting, about a month after it adopted the resolution noted above, H. Lee Dennison had been appointed to the council. At that meeting he expressed his own views favoring completion of the Corps of Engineers project at Westhampton Beach by the construction of six more groins. The council, therefore, was faced with the policy problem of applying its general position in the resolution to the specific beach erosion situation at Westhampton. The council decided at that time not to amend its resolution, but rather to appoint a committee to examine this specific problem. The findings of this investigation could either strengthen the resolution or be expressed as an exception to it. The chairman appointed Barry D. Andres, H. Lee Dennison, and J. R. Welker to this special committee.[14]

The special committee on the shoreline met on June 5, 1973. In spite of the differences between Mr. Dennison and the other members of the committee, they came to agreement on the following points:

1. Ideally there should be public ownership or public interest in the shore. Lease arrangements could be made with residents.
2. There should be sand bypassing systems at the inlets.
3. Washovers will not harm wetlands or marine life in the bays.
4. As a *general* rule, groins and jetties should not be built.

As might be expected, there was disagreement about groin construction at Westhampton Beach. Dennison favored construction of the six groins to complete the project. Andres and Welker opposed building the groins because they believed the groins would not protect the beach or prevent washovers.

Mr. Dennison had to leave the meeting early. The committee report noted that further discussion produced the following alternatives: "If it is determined that the County is legally responsible for the erosion west of the last groin, either complete the project or buy the land west of the last groin; let a new inlet form; close Moriches [Inlet] and effect sand bypassing at the new inlet."[15] The final conclusion of the report was that an environmental impact statement should be prepared on the proposed completion of the Westhampton project. It should clearly define benefits to the public and include cost/benefit analyses regarding recreation, boating, shellfishing, and protection of the mainland. Annual maintenance costs, alternatives to groin construction, and various methods of sand bypassing should also be detailed in the statement.

The committee report was presented at the July 24, 1973, meeting of the council. The committee noted that it did not have sufficient data to make definite recommendations about the proposed Westhampton groins, and therefore suggested the preparation of an environmental impact analysis. The full council discussed the committee report at length, and many differences of opinion were expressed. County Legislator H. Beecher Halsey stressed the point that existing extensive development at Westhampton foreclosed the possibility of returning to reliance on natural beach dynamics. The council unanimously agreed that it would not try to reach a decision at that meeting. It would wait until the environmental impact statement and a study being made for the Corps of Engineers by Greenman, Pederson and Associates were completed.[16]

FOOTNOTES

[1] Resolution No. 111-1967.

[2] Resolution No. 112-1967.

[3] Resolutions No. 187-1968 and No. 264-1968.

[4] Resolution No. 467-1968, Minutes of the Regular Meeting of August 12, 1968, and Resolution No. 635-1968.

[5] Information obtained from Paul Fox, Office of the County Treasurer.

[6] County of Suffolk Capital Program for 1971-1973, p. 61.

[7] Resolution No. 541-1970.

[8] County of Suffolk Capital Program for 1972-1974, p. *x*.

[9] *Ibid.*

[10] County of Suffolk Capital Program for 1973-1975, pp. 4-5.

[11] Resolution No. 539-1972, referring to Project Number 5317.

[12] Interview with DeWitt Davies, staff member of the Regional Marine Resources Council, June 18, 1974.

[13] Bagg, James F. "An Evaluation of Barrier Beach Dynamics," May 1973.

[14] Suffolk County Council on Environmental Quality, Minutes of May 23, 1973 Meeting, pp. 4-5.

[15] Memorandum. "Summary of Special Committee on the Shoreline, Interim Report, Meeting of June 5, 1973," Council on Environmental Quality.

[16] Minutes of July 24, 1973 Meeting, Council on Environmental Quality, p. 5.

CHAPTER 8

LOCAL GOVERNMENT IN BEACH EROSION CONTROL

The Town of Southampton and the Village of Westhampton Beach are the two local governments most concerned about the beach erosion problem. The first groins were constructed on that portion of the Moriches to Shinnecock barrier beach that is within the boundaries of the village. There is no longer an erosion problem there, for the first stages of the Corps of Engineers beach stabilization and hurricane protection project have apparently succeeded in maintaining and even enlarging the beach and dunes. The Town of Southampton, however, must still deal with the consequences of continued erosion at its Neighborhood Recreation Center beach and private properties nearby.

The town does not participate financially in the beach stabilization and hurricane protection project. It must therefore depend on the consensus of federal, state, and county governments for decision about whether or not to complete the groin field. Nevertheless, beach fill, dune reconstruction, and groins are not the only measures necessary to maintain the fragile shoreline. Human action may be as destructive as winds and waves, especially on the dunes, which are the second line of defense (after the beach) against flood damage caused by storms. Controls over building, vehicle use, and protection of the vegetative cover are important contributions of local government to shore stabilization.

TOWN OF SOUTHAMPTON

The beach erosion problem at the westerly end of Westhampton Beach is within the territorial jurisdiction of the Town of Southampton. Erosion and storm damage washed away the town beach, officially the Neighborhood Recreation Center at Westhampton Beach, and the parking lot on the south side of Dune Road. This is a direct financial loss, but the town is also concerned about the erosion and damage to private property near the town beach. Theodore Hulse, the town supervisor, notes that these residences are occupied mostly in the summer. Their owners send almost no children to school, but they make a substantial tax contribution to the school district. They also require few services from the town in exchange for property taxes paid.[1]

The Town of Southampton has a planning board, but the organization has no technical staff of its own. Both the Town Board and the Planning Board depend for technical information about the beach erosion problem on the town's consulting engineers, Greenman, Pederson and Associates. The consulting engineers note that expert opinion supports the theory that the existing groins east of the beach washout have aggravated beach erosion, although this was a serious problem long before the groins were constructed. In any case, if the groins have aggravated the problem, it is because they are too long.[2]

It has already been noted that the state and county provide the full nonfederal share of financing for beach erosion control and hurricane protection. Because it has no financial interest in the project, the town does not officially participate in decision-making. Nevertheless, the Town Board of the Town of Southampton has adopted the policy position that it favors the completion of the groin field originally planned by the Corps of Engineers, which was to have extended westward beyond the town beach. The Town Board resolution adopted May 1, 1973, and transmitted to Suffolk County Executive John V. N. Klein, concluded that the Board

> Hereby Resolves, to prevail upon the County of Suffolk, the Department of Transportation, the Department of Environmental Conservation of the State of New York and the Army Corps of

> Engineers of the United States to complete the groin field at Westhampton, Town of Southampton, and to proceed with utmost haste in developing and implementing an erosion control and hurricane protection project for the remainder of our barrier beach to preclude the possibility of an imminent major disaster, with huge property losses and probable loss of life.[3]

Supervisor Hulse is personally convinced of the success of the Corps of Engineers beach erosion control design. He is familiar at first hand with the effectiveness of groins in stabilizing and even widening the beach sections between them. At one time he managed property on the Westhampton Beach and observed how the beach widened over the years after the groins were completed. This observation was confirmed from the air when he flew over the beach with a group of Suffolk County legislators. They noted the obvious contrast between the protected sections of the beach and the areas both east and west of the groin field.[4]

Government of the Town of Southampton

Local government on Long Island retains some colonial features along with its contemporary structure as mandated by the Town Law of the State of New York. The Town Board is the governing body of the town according to the state statute, but its jurisdiction over town lands is limited. Absolute ownership is vested in the Board of Trustees of the Freeholders and Commonalty of the Town of Southampton. Because its powers were derived directly from King James II, its sovereignty over its areas of jurisdiction as set forth in its charter, or patent, are considered to be superior even to that of the State of New York. The Town of Southampton has two royal charters. The Andros Charter was dated November 1, 1676; the Dongan Charter was granted under royal authority by the Governor of the Province of New York on December 6, 1686. This is the famous Dongan Patent that assigned ownership and jurisdiction over all undivided lands, all waters, all lands under water, and the produce of these lands and waters to the Board of Trustees of the Freeholders and the Commonalty of the Town of Southampton and their heirs, successors, and assignees forever.[5]

In 1818 and 1831 the legislature of the State of New York reaffirmed the rights and privileges granted by the Dongan Patent. At the turn of the century a court decision further supported the claim of the Board of Trustees in People *ex rel.* Howell *v.* Jessup (160 N.Y. 249). In 1932 the Appellate Division reaffirmed this finding in the "Southampton Town Clam Case,"[6] as it is called.

The town Neighborhood Recreation Center at Westhampton Beach includes beaches on both the Atlantic Ocean and Moriches Bay shores of the barrier island as well as Dune Road and the adjacent parking lots. Further erosion and breaching of the barrier island at this location would not only destroy this town property, but also affect the shellfish industry and fisheries in the town waters of Moriches Bay, the tidal wetlands, and public and private lands on the north shore of the Bay. For this reason, the Board of Trustees has supported efforts to complete the beach stabilization and hurricane protection project by the construction of the last six groins. This position was expressed in a letter to County Representative H. Beecher Halsey, Jr.: "It is the unanimous opinion of this Board that the completion of the groin field is a necessity. It is our feeling that the property of the Freeholders of Southampton Town, Moriches Bay, would be protected by the groins and from future blow-outs."[7]

The authority of the Board of Trustees has lately been challenged by the state. The Tidal Wetlands Act of 1973 requires that permits be obtained from the Department of Environmental Conservation for any alteration of wetlands by filling or building piers, bulkheads or other structures. Before the advent of this act, the Board of Trustees had sole jurisdiction over town wetlands and waters, granting such permits according to its own criteria. Now the state, through DEC, is attempting to override the authority of the Board of Trustees. The Department does not dispute the Board of Trustees' ownership of the tidal wetlands. It claims that the Board has only the rights of any other property owner. The state is sovereign and has the same jurisdiction to regulate Board of Trustees' land as it has over other lands. Applications to alter wetlands by town residents or the town itself should be submitted to and granted by the state.

The Town of Southampton strenuously objects to this attempted invasion of its own sovereignty under the Andros and Dongan Patents. Together with the Towns of Easthampton and Southold it will sponsor legislation to exempt the three towns from the Tidal Wetlands Act. If this fails, they will defend their rights in court.[8]

The structural uniqueness of the Board of Trustees and conflict with the state agency point up the political isolation of Long Island from the seat of state government in Albany. Contacts between town and state agencies are rare. There has been little exchange of views or information between town and state on coastal zone matters, including the Westhampton Beach erosion problem. The only real communication between the town and Albany is through its representatives in the legislature. Southampton is fortunate in having as one of these Perry Duryea, the powerful former Speaker of the State Assembly.

Nevertheless, new emphasis on coastal zone planning and management will also require expanded communications between the responsible state agencies, which are the state planning office in the Department of State and the Department of Environmental Conservation, and local planning and governing bodies. It is likely, however, that such contacts will be filtered through another organization unique to Long Island, the Nassau-Suffolk Regional Planning Board. Multicounty regional planning boards exist in other parts of the state, but none seems to have the influence of the Long Island body and its director, Dr. Lee E. Koppelman. Nassau and Suffolk County Executives have agreed to finance a $3 million study by the Board of the Great South Bay. Dr. Koppelman also anticipates preparing the entire Long Island segment of the New York State coastal zone plan, although preparing any statewide plan would be the responsibility of the Secretary of State.

The Town of Southampton Master Plan

The predominant position of the Nassau-Suffolk Regional Planning Board tends to distract attention from the statutory planning responsibilities of local governments. The Nassau-Suffolk Regional Planning Board has properly focused its attention on regionwide problems of land use, transportation, utilities, and environment. Long Island cities, towns, and villages must prepare their own development plans; and it is on the basis of these plans that actual zoning and land-use regulations are formulated and enforced according to local perceptions. The mayor of one village was impressed with the wide shelf of volumes on planning made available by the Regional Planning Board but found little in this material that was helpful in dealing with purely local problems. The director of a town planning agency, who was also well aware of the high quality of the work of the Regional Board, found some lack of understanding and sympathy with local affairs.[9]

The Town of Southampton Master Plan, prepared by a consulting firm with federal financial aid, was published in 1970. It recommended this policy with regard to dune land and ocean beaches: "Both dune land and ocean beaches should be protected from any improvement or use that would tend to limit their effectiveness as natural barriers to ocean wave overruns. No development or improvement, other than an approved beach protection measure or an approved walk over the dune area, shall be constructed closer than 40 feet inland of the natural crest of the foreward dune."[10]

This policy was implemented by the adoption of revised Building Zone Ordinance Number 26 on May 2, 1972. The reach of Westhampton Beach in which erosion has taken place is zoned R-40 Residence. Permitted uses are single family detached dwellings on lots with a minimum of 40,000 square feet; maximum permitted lot coverage is 20%. This area is also covered by two overlay districts that impose restrictions more stringent than in other R-40 districts. These are the Tidal Wetland and Beach Overlay District and the Tidal Flood Plain Overlay District.

Tidal Wetland and Ocean Beach Overlay District

Section 2-40-10.01 of the Building Zone Ordinance sets forth the purpose of this district as the protection of "a unique and relatively scarce environment" necessary for propagation of fish and shellfish, important for commercial and sport fishing, for waterfowl habitat, and for the scenic character that supports recreational and tourist activities. "Therefore, the Town enacts these provisions as a matter of public policy, recognizing the extremely insecure status of these valuable natural resources, particularly in a time of rapid technological progress and urbanization, their limited suitability for development and the interest of all its people in these geographic areas, especially as established in the Dongan Patent."

With reference to the ocean beach the overlay district provisions prohibit building on the public beach except on approval by public agencies and then only to protect the natural environment or for some other public purpose. Owners of private lands are prohibited from excavating, regrading, or disturbing the natural crest of the dune, "except in accordance with a beach protection policy or protective works program approved by the Town."[11] Furthermore: "A private landowner shall not construct any building or structure on the ocean beach except that one (1) access walkway shall be constructed over the crest of the dune provided that it be established at an elevation of at least two (2) feet above the undisturbed natural crest where grasses exist or directly on the grade without any space between where no grasses exist."[12]

Board walkways or stiles over the dune crest to the beach, even from one's private home, are required to protect the dune from erosion. Unrestricted walking on the dunes could disturb the vegetation and dune profile. In the Village of Westhampton Beach, large signs announce the prohibition, by ordinance, of walking on the dunes. The dunes are further protected by provisions of the Tidal Flood Plain Overlay District.

Tidal Flood Plain Overlay District

The importance of the dunes for defense of the barrier island and the back bay uplands from storm surge flooding is noted in the stated purpose of the district: "Since the recorded history of Atlantic Coastal storms and hurricanes establishes the fact that flooding of the ocean beach and uplands along the bays can be anticipated with consequent danger to life and health as well as property damage and other related hazards . . . Southampton hereby establishes a Tidal Flood Plain Overlay District derived by those elevations of land that approximate the standard project Design Hurricane."[13]

To accomplish this purpose the Building Zone Ordinance requires that on the entire barrier beach all dwellings must have a first floor elevation at least 15 feet above mean sea level. The original ordinance required a building setback of 50 feet shoreward from the crest of the first rank of dunes, but this was increased to 100 feet in a later amendment. On shallow lots, required frontyard setbacks may be reduced by 50% to accommodate the setback from the dunes. If the dune line along any stretch of beach is not readily discernible, an application for a building permit must be treated as a Special Exception Use and be reviewed by the Board of Appeals.[14]

Other Dune Protection Ordinances

Beach Protection Ordinance Number 8, adopted in 1939, prohibits disturbance to any "revetment, fence, work, installation of any part thereof, or any beach grass," which has the purpose of aiding in the restoration of the dunes or natural sand barrier on the south beach of the town. Ordinance No. 14, which was adopted in 1945, prohibits the removal of any grass or natural growth, except poison ivy, from beaches and dunes in the town except by permit and under appropriate conditions.

Vehicles such as dune buggies and jeeps are especially destructive of dunes. Autos and other vehicles are prohibited from operating on the dunes by Ordinance No. 22, originally adopted in 1953. A 1959 amendment permits the use of beach buggies on the beaches, not the dunes, outside the incorporated villages, between the hours of 6:00 p.m. and 10:00 a.m. from June 15 through September 15.

Although the Town of Southampton does not participate directly in the cooperative beach stabilization and hurricane project designed by the Corps of Engineers, it does play an important role in protecting the beaches and dunes. Before local zoning, land use, and beach protection ordinances were adopted, homes and other structures were erected on the dune crest, and the dunes were sometimes demolished to open the view or provide access to the beach. Dune reconstruction and beach fill and stabilization by groins would be to little avail if the density and conditions of development on the barrier beach were not regulated. This is a local government responsibility.

VILLAGE OF WESTHAMPTON BEACH

The entire barrier island between Moriches and Shinnecock Inlets has been referred to as Westhampton Beach, although it falls into several local jurisdictions. The western tip of the island is in the Town of Brookhaven. The central part is in the Village of Westhampton Beach. The remainder of the island on either side of the village boundaries is in the Town of Southampton. The town Neighborhood Recreation Center Beach (and its erosion problem) is to the west of the village.

The Corps of Engineers' survey report for the Moriches to Shinnecock beach stabilization and hurricane protection project was cautious in its recommendations about the construction of groins. The report stated they might possibly be included; but it noted that the consequences of groin construction were not entirely predictable because there was not enough experience with groins on Long Island ocean beaches. It also warned that groin construction should be started, if at all, at the western end of the Moriches to Shinnecock island. If they were begun elsewhere, care should be taken to fill the new beach out to the ends of the groins, so that they would not trap sand from the littoral current and deprive the beaches to the west. Neither of these conditions was fulfilled, as earlier chapters have related.

In 1963 severe storms caused critical beach erosion in the Westhampton Beach Village section of the barrier island. The groin construction and beach fill were therefore started in the middle of the island within the village. Eleven groins were built, and then Suffolk County halted the project by failing to appropriate further matching funds. Erosion, aggravated by storm damage, continued to the west of this groin field, so four additional groins were authorized in 1969. Construction was completed in November 1970, but beach fill was not placed out to the end of these groins.

Although the section of beach to the west of the last four groins is not in the Westhampton Beach Village jurisdiction, the village did take a position favoring construction of the last six groins to protect the beach between the existing groins and Moriches Inlet. This support was expressed in a letter from Theodore O. Hulse, then mayor of the village, to H. Beecher Halsey, Jr., representative to the Suffolk County Legislature. The mayor urged that Suffolk County should proceed with groin construction to complete the federally designed beach erosion control and hurricane protection project in the Town of Southampton. The expenditure involved would not simply be a large public investment to protect private property, as the County Executive claimed. The county itself owned property worth over $3 million between the eroded beach and Moriches Inlet. He summarized the public property and income values in 1970 as follows:[15]

Suffolk County Property	
3 miles of County road	$ 180,000
County park with 6,400 feet of beach frontage	2,560,000
County park facilities and parking lot	160,000
Water mains, meters, fire hydrants - 14,555 feet	218,000
	$3,118,300
Town of Southampton Property	
Town beach with 300 feet of beach frontage	120,000
Total property value	$3,238,000
Taxes and Fees Paid by Residents	
To County of Suffolk	55,234.87
To Town of Southampton	35,790.18
To School District	103,097.52
To Village of Westhampton Beach	4,088.00
To Suffolk County Water Authority	11,710.87

The village government has not altered its position since this letter was written. The present mayor, C. E. Moore, has stated his position as favoring either completion of the groin field or removing the existing groins.[16] The situation is critical because the existing groins appear to be depriving the beach to the west of sand nourishment. Mayor Moore believed this situation should not be allowed to continue.

Beach and Dune Protection Ordinances

In recognition of the vital role played by dunes in protecting barrier islands and back bays from storm surge flooding, the Village of Westhampton Beach has enacted two dune-protection ordinances. One has the title "Moving of Sand on the Dune" and is intended to prevent sand mining. Permits from the village Superintendent of Public Works are required before sand may be moved, but "in no event shall a permit be issued or granted if the movement of sand proposed is: (a) from the property on the south side of Dune Road to property on the north side; (b) from the barrier beach to some portion of the mainland."

The other ordinance is concerned with "Building and Structure on the Dunes." It generally prohibits any action that would damage the dunes or snow fences, grass or other measures intended to stabilize the dunes. Even movement on the dunes is prohibited: "Section 3. No person, firm or corporation shall walk on, over, or along, or drive any animal or vehicle up, over or along the sand dune or any portion of the north or south slope thereof, except by the use of a stile, steps, walkway or other device which is sufficient to protect the sand slopes or sand dune from any damage or change by reason of such use."[17]

Building on the dune itself is prohibited, and location and structure are regulated to avoid damage to the dune and reduce damage to the structure from possible storm surge flooding. The barrier island is zoned R-3, single-family homes on 20,000 square-foot lots. No house may be built closer than 35 feet from the crest of the dune; and in that location the bottom floor joists must be at least 20 feet above mean sea level. Lower elevations are permitted further back from the dune line, but floors must be at least 10 feet above mean sea level.

In order that flood waters may pass under the houses, they must be built on piles driven and spaced according to specific requirements. The space under the first floor may be enclosed only by wood walls of light construction, or their equivalent as specified by a licensed professional engineer. These would presumably provide no obstacle to the passage of flood waters under the house.

FOOTNOTES

[1] Telephone interview with Theodore Hulse, Supervisor, Town of Southamp- tom, January 2, 1974.

[2] Telephone interview with Robert Grover of Greenman, Pederson, and Associates, January 2, 1974.

[3] Letter from Claude Jones, Town Clerk, to John V. N. Klein, Suffolk County Executive, dated May 4, 1973.

[4] Interview with Supervisor Hulse, June 5, 1974 in the Town Hall, Village of Southampton.

[5] Meschutt, Stephen F., Ed. *The Board of Trustees of the Freeholders and Commonalty of the Town of Southampton* (1968), pp. 8-17.

[6] People of the State of New York *v.* Richard Miller and the Trustees of the Freeholders and Commonalty of the Town of Southampton, 235 AD 226, April 29, 1932.

[7] Letter dated March 6, 1974 signed by Thomas F. Rewinski, Acting Chairman.

[8] Remarks by members of the Board of Trustees of the Freeholders and Commonalty of the Town of Southampton at its meeting on June 24, 1974, at the Town Hall.

[9] Personal interviews with author.

[10] Town of Southampton Master Plan, 1970, p. 93.

[11] Section 2-40-10.03 (b).

[12] Section 2-40-10.03 (1).

[13] Section 2-40-20.01.

[14] Section 2-40-20.02 (a) and 2-40-20.04 (a) and (c).

[15] Letter from Theodore O. Hulse, Mayor of Village of Westhampton Beach to Rep. H. Beecher Halsey, Jr., dated January 7, 1970.

[16] Telephone interview with C. E. Moore, Mayor of Village of Westhampton Beach, October 11, 1973.

[17] Building and Structure on the Dunes, Ordinances of the Village of Westhampton Beach, June 29, 1970.

CHAPTER 9

INSTITUTIONAL STRUCTURE FOR COASTAL ZONE MANAGEMENT

Coastal zone planning and management is one component of the rivalry between state and local governments for control of environmental resources and regulation of land use and development. Under the federal system of government these powers are constitutionally reserved to the states. The states, however, have delegated these powers, to one degree or another, to local governments and have sometimes reinforced this arrangement with a legal structure for local home rule. Having once acquired jurisdiction over land use and development, local governments are reluctant to relinquish it back to the states.

In the days when communities were comparatively small and isolated, the tradition of local autonomy had many advantages. As they grew and coalesced into large urban areas and as transportation technology increased, the accessibility of natural resource and recreation areas to large numbers of people (and, consequently, the problem of how such areas should be used and developed) became matters of regional and statewide concern. This is an important current political problem. Shall local governments alone determine how land within their boundaries is to be used, according to their own local interests, or should the state reassert its constitutional power to review local determinations or itself assume jurisdiction where land use and natural resources conservation are of compelling wider public interest?

Regardless of the possible merits of recovery by the state of its constitutional powers over local affairs, as a practical matter it is beset by political obstacles. Recentralization of power in the state capital would require legislative action to repeal or modify the statutes that delegated land-use control and other development regulation powers to the local governments. State legislators elected by local constituencies opposed to giving up such powers are not likely to advocate any weakening of the structure of home rule. Just before the November 1974 elections, the Long Island newspaper *Newsday* looked into this aspect of state-local relations. An editorial reported as follows:

> In our poll of legislative candidates, we asked whether they thought the state or the counties should be given the power to oversee local planning and zoning decisions. . . . They didn't [think so] because too many candidates cling to the home rule formula. At the last session of the Legislature, for example, State Senator Bernard Smith introduced a bill to set up a [state] review for land-use decisions in the coastal zones. . . . The bill was given short shrift because it threatened the home rule powers of the towns and villages.

Four issues of state-local relations will have to be resolved if New York State is to carry out a successful coastal-zone planning and management process. *First*, how to restore for the state its constitutional powers over local government in order to conserve natural resources and other areas of vital statewide concern. *Second,* how to organize the planning process so that it will take account of such statewide interests and yet formulate a coastal zone plan that will reflect the planning efforts and concerns of local governments. *Third*, how to organize a procedure for the review of local planning and development control decisions to assure that they will conform to the statewide coastal zone policy framework agreed to in the adopted state plan. *Fourth*, how to coordinate state coastal zone planning and management with other aspects of environmental resource management and state development policies in general. These issues will be examined here in light of the existing state structure for planning and land use control and alternatives that have been suggested for changes.

EXISTING STATE STRUCTURE FOR LOCAL ENVIRONMENTAL PLANNING

Four state statutes govern local planning and land use regulation. They are the General City Law, the Town Law, the Village Law, and the General Municipal Law. Towns must adhere in these matters to the provisions of the Town Law. Cities and villages, however, may organize these activities under either the City or Village Laws or the General Municipal Law. The latter also authorizes planning by counties and metropolitan or regional agencies.

Local Planning Boards and Commissions

Under the General City, Town, and Village Laws, the local governing bodies may establish planning boards and appropriate funds to support their activities. Their planning function is defined in language similar to Section 28-a of the General City Law:

> The planning board may prepare and change a comprehensive master plan for the development of the entire area of the city, which master plan shall show existing and proposed streets, bridges and tunnels and the approaches thereto, viaducts, parks, public reservations, roadways in parks, sites for public buildings and structures, zoning districts, pierhead and bulkhead lines, waterways and routes of public utilities and such other features existing and proposed as will provide for the improvement of the city and its future growth, protection, and development, and will afford adequate facilities for the public housing, transportation, distribution, comfort, convenience, public health, safety and general welfare of its population.

In addition to preparing the master plan for the community, the local planning board may also be authorized to participate in development policy issues. The local governing body may require "the reference of any matter or class of matters to the planning board" before other officers or bodies having final authority in these matters may take action on them.[1] For example, under such a rule the planning board would have authority to review and report to the governing body on plans for streets, utilities, or waterfront projects before the city engineer could proceed to carry out construction.

Instead of creating a planning board, cities and villages may establish planning commissions under Article 12-A of the General Municipal Law. This law states, "This article shall be construed as the grant of additional power and authority to cities and incorporated villages."[2] This additional power is the mandatory referral to the planning commission of the following matters before action may be taken on them by the public officers having final jurisdiction:

> The adoption of any map or plan of said city or incorporated village, or part thereof including drainage and sewer or water system plans or maps, and plans or maps for any public water front, or marginal street, or public structure upon, or in connection with such front or streets, or for any dredging, filling or fixing of lines with relation to said front; or any changes of any such maps or plans; the location of any public structure upon, in or in connection with, or fixing lines with relation to said front; the location of any public building, bridges, statue or monument, highway, park, parkway, square, playground or recreational ground, or public open place of said city or village.[3]

In addition to such mandatory referrals, planning commissions would have the master plan preparation responsibility authorized by the General City and Village Laws. Section 237 of Article 12-A of the General Municipal Law also provides for the following planning responsibilities:

> Such planning commission may cause to be made a map or maps of said city or village or any portion thereof, or of any land outside the limits of said city or village so near or so related thereto that in the opinion of said planning commission it should be so mapped. Such plans may show not only such matters as by law have been or may be referred to the planning commission, but also any and all matters and things with relation to the plan of said city or village which to said planning commission seems necessary and proper, including recommendations and changes suggested by it.

Planning board and commission reports are only advisory to the governing body. Master plans, even when adopted by local governments, have no legal or binding effect. If a community wishes to implement its land use and development plans, it must use legal tools such as zoning, subdivision regulations, and the official map.

Zoning

Cities, towns, and villages have somewhat differently worded authorization for zoning, the primary land use control. Section 20 of the General City Law states that every city is empowered:

> To regulate and limit the height, bulk, and location of buildings hereafter erected, to regulate and determine the area of yards, courts and other open spaces, and to regulate the density of population in any given area, and for said purposes to divide the city into districts. . . .
>
> To regulate and restrict the location of trades and industries and the location of buildings, designed for specific uses, and for said purposes to divide the city into districts and to prescribe for each such district the trades and industries that shall be excluded or subjected to special regulation and the uses for which buildings may not be erected or altered.

Town boards and village boards of trustees are also empowered by ordinance or local law "to regulate and restrict the height, number of stories and size of buildings and other structures, the percentage of lot that may be occupied, the size of yards, courts and other open spaces, the density of population, and the location and use of buildings, structures, and land for trade, industry, residence or other purposes."[4]

The Village Law has the unique provision that before the board of trustees may act on a zoning ordinance for the first time, it must appoint a zoning commission to recommend the boundaries of the districts and the regulations to be enforced in them. The zoning commission must make a preliminary report on a zoning ordinance, hold public hearings on it, and then submit a final report before the village trustees may hold their own hearings or take action.[5]

Subdivision Control

Under the state enabling laws for local planning, municipal governing bodies that have created planning boards may by ordinance or resolution authorize the planning board to approve land subdivision plats showing lots, blocks or sites, with or without streets or highways. When required by the planning board, the plat must include suitable parks and playgrounds.

When this is not a practical requirement for the site, the governing body may fix the amount of money the developer shall pay as a contribution to a park trust fund. The subdivision regulations contain standards for streets, which should conform to the official map and proposals shown on the master plan. Of particular relevance to the coastal zone is the requirement that "the land shown on such plats shall be of such a character that it can be used safely for building purposes without danger to health or peril from fire, flood, or other menace."[6]

The land and building uses in approved subdivision plats must conform to the local zoning ordinance. It is possible, however, for a developer to propose a planned unit development that may have a mixture of housing types and densities and include nonresidential uses. After the planning board holds a hearing on the proposal, it may recommend to the local governing body an amendment to the zoning ordinance that would incorporate the planned unit development.

Although the local legislative body is the final authority on adoption and changes in zoning, the planning board is the final authority for approving subdivision plats. This is a considerable grant of power by the state statutes, for "no plat of a subdivision of land . . . shall be filed or recorded in the office of the county clerk or register until it has been approved by a planning board."[7] The importance of this power to the developer is that he would be practically unable to sell lots or record deeds for a plat that had not been recorded by the county. Without planning board approval of the plat, the streets could not be dedicated to the municipality for future maintenance, snow plowing, and repair.

Official Maps

Local legislative bodies have a third legal tool for planning and managing their environmental development. This is the official map, which shows the locations and boundaries of existing and future streets, highways, parks, and drainage systems. Because it shows the rights-of-way of future streets and the other public facilities, the official map has a considerable impact on private property. The owners are unable to obtain a permit for a building in the bed of a mapped street. In addition, they may not obtain building permits "unless a street or highway giving access to such proposed structure has been duly placed on the official map or plan, which street or highway shall have been suitably improved to the satisfaction of the planning board."[8] Furthermore, no public municipal utility or improvement may be constructed in any street or highway until it has been included in the official map.

Appeals

No development regulation, whether it be zoning, subdivision control, or the official map, can foresee its consequences on every parcel of property in the community. There may be situations in which the characteristics of the property, its shape or topography for example, do not lend themselves to economic use in strict conformity with legal requirements. Local legislative bodies may, therefore, establish boards of appeals. Property owners who feel they are aggrieved by the decision of a planning board or permit-granting officer in connection with any of these legal development controls may go before the board of appeals. After hearing, the appeals board may vary the strict letter of the local regulation to accommodate a particular hardship or practical difficulty.

The local appeals board does not have the last word, however. If they are still not satisfied, property owners may apply for relief to the Supreme Court in their county under Article 78 of the Civil Practice Law. Once in the courts, cases may be appealed to higher levels. Zoning cases have even been heard and decided in the U.S. Supreme Court.

COUNTY, METROPOLITAN, AND REGIONAL PLANNING BOARDS

Article 12-B of the General Municipal Law is the state authorization for the establishment of county, regional, and metropolitan planning agencies. County legislative bodies may establish county planning boards. Counties may also join with municipalities and other counties to form regional and metropolitan planning boards in which all the component governments are represented. State law has given these planning agencies, where they have been established, considerable technical planning responsibilities and some limited powers to review certain municipal planning and zoning decisions. In one case, Suffolk County, the planning board has the power to veto these particular zoning determinations.

There are two reasons for summarizing local, county, and regional (hereafter including metropolitan) planning powers and responsibilities in such detail. First, as already noted, the existing institutional structure that has been established by state law would have to be dismantled or changed for a different structure to be erected. Whatever the merits of a proposed change, there would be considerable political obstacles in the way of inducing local governments to give up powers already vested in them. Second, one alternative to changing the state structure of land use and development controls is to develop the existing structure by strengthening the roles of county governments and regional planning agencies. This will be discussed in Chapter 10.

County and regional planning boards are authorized to carry out planning studies and prepare and adopt comprehensive master plans for the entire area covered by the constituent counties and municipalities. These are only advisory documents, however, unless adopted by a county legislative body. Such a county plan would show "the highways, parks, parkways and sites for public buildings or works . . . other than state or federal projects, in the acquisition, financing or construction of which the county has participated or may be called upon to participate."[9]

The extraordinary feature of county plans, once adopted after proper public hearing, is that they become legally binding on the county government.

> When so approved in whole or in part by the board of supervisors in any county or counties such approved county plan or part thereof shall be deemed to be binding upon the board of supervisors of the county and the several county departments thereof, and no expenditure of public funds by such county for the acquisition of land except for state or federal projects

> or for any public improvements shown on such county plan shall be made except in accordance with such county plan.[10]

If the county legislature wanted to finance public improvements not shown on the county plan or that would require changes in the plan, the plan would have to be amended after review and advice by the county planning board and a public hearing.

Note that the legally binding county plan covers only public works or public improvements, not land use. Zoning powers are reserved to municipalities by the state laws already described. County or regional planning boards may, however, analyze land uses and "recommend to the governing bodies of the several cities, villages, and towns . . . a comprehensive zoning plan."[11] Furthermore, municipal planning and zoning actions affecting areas within 500 feet of the boundaries of neighboring communities or certain public facilities may be reviewed by county or regional planning boards.

> It is hereby declared that it is in the public interest that certain classes of zoning and planning actions by a city, town, or village . . . be reviewed by the county planning agency . . . or in the absence of such a county planning agency, by the metropolitan or regional planning agency, if any . . . as an aid in coordinating such zoning actions and planning among municipalities by bringing pertinent intercommunity and countywide considerations to the attention of the aforesaid municipal agencies having jurisdiction.[12]
>
> Each municipal body which has jurisdiction to adopt or amend zoning regulations, or to issue special permits or grant variances pursuant to such regulations, shall before taking final action on certain of such matters, refer the same to such county, metropolitan or regional planning agency.
>
> The matters covered by this section shall include: (a) any municipal zoning regulation, or any amendment thereof, which would change the district classification of or the regulations applying to real property lying within a distance of five hundred feet from the boundary of any city, village, or town, or from the boundary of any existing or proposed county or state park or other recreation area, or from the right-of-way of any existing or proposed county or state parkway, thruway, expressway, road or highway, or from the existing or proposed right-of-way of any stream or drainage channel owned by the county or for which the county has established channel lines, or from the

> existing or proposed boundary of any county or state owned land on which a public building or institution is situated; and (b) any special permit or variance affecting such real property within such distance of 500 feet.[13]

County or regional planning agency review of local zoning decisions is advisory only and not binding on the local governments (with one exception, Suffolk County). Nevertheless, if the county or regional body disapproves or recommends modification of a proposed local action, "the municipal agency having jurisdiction shall not act contrary to such disapproval or recommendation except by a vote of a majority plus one of all the members thereof and after the adoption of a resolution fully setting forth the reasons for such contrary action."[14]

The unique case of Suffolk County with regard to review of local zoning determinations should be considered in some detail. The fact that the county charter takes precedence over a general state law is interesting enough, but it also furnishes a lead toward developing an institutional structure for higher level control over local land use and development planning in the coastal zone. This point will be elaborated in the next chapter. What will be noted here is the existing legal basis for Suffolk County Planning Commission power to veto certain local zoning decisions.

In the November 1970 election, Suffolk County voters approved an amendment to the County Charter. Section 1330 of the Charter now reads as follows:

> Power to zone within five hundred feet of town or village boundary restricted. In addition to the other provisions of this article concerning municipal zoning actions, no zoning ordinance nor any amendment of a zoning ordinance passed by any town or village in the county relating to any portion of the said town or village within five hundred feet of a town or village boundary shall take effect in respect to such portion of said town or village until said ordinance or amendment has been submitted to and approved by the county planning commission.

If there is an objection by the neighboring municipality affected and, after a public hearing, the County Planning Commission disapproves of the proposed zoning ordinance or change by a two-thirds vote, the proposal would in effect be vetoed.

This extraordinary power of the Suffolk County Planning Commission was tested in the New York Court of Appeals in the case of Town of Smithtown *v.* Ralph D. Howell, Sr., *et al.*, Constituting the Suffolk County Planning Commission.[15] The legal action stemmed from the approval by the Town of Smithtown of a zoning change from "quarter-acre residential" to "wholesale and service" to permit the establishment of a franchised auto dealer on a parcel of property within 500 feet of the boundary of the Village of The Branch. Zoning of the adjacent area in the village was residential. The Suffolk County Planning Commission vetoed the zoning change. The Town of Smithtown brought suit against the County Commission. When the case reached the Court of Appeals, the highest in the state, the court decided that under the state constitution, Home Rule Law, and the Suffolk County Charter, certain zoning powers had been transferred from the towns to the county. The County Planning Commission therefore had the legal power to veto town zoning decisions within 500 feet of town boundaries.

Article IX, Section 1 of the state constitution empowers counties to adopt, amend, or repeal alternative forms of county government. Counties may also, with a double referendum, transfer functions between cities, towns, and villages and the county. Such a charter provision must be approved, however, by a majority vote in both the area of the county outside of cities and villages, and in the cities, if any, considered as one unit and the villages, if any, considered as one unit.

Section 33, subdivision 7a of the Home Rule Law implements this constitutional grant of power to counties to transfer functions. Article 1330 of the Suffolk County Charter was enacted in accordance with this section of the Home Rule Law. The court found, therefore, that "the rather stringent requirement that there be a direct voice of the electors residing in the several kinds of localities affected, explains the otherwise unusual grant of power to counties to transfer functions from one layer of government to another." The County Charter, in effect, transferred final determination of zoning changes within 500 feet of a municipal boundary from local governments to the County Planning Commission. What is even more unusual is that this section of the County Charter supersedes a state law, the General Municipal Law. Section 239-m of that law authorizes only advisory review powers over local zoning to county planning agencies, not veto power. The Suffolk County Charter amendment was approved by the voters after the enactment of the latest version of the General Municipal Law. The Court of Appeals, therefore, found that the Suffolk County Planning Board had zoning review powers beyond those granted by general state law.

Although county and regional planning boards do not have zoning powers, they may exercise subdivision control outside the limits of any incorporated village or the area over which a city exercises plat approval powers. The county or regional planning board would adopt subdivision regulations in accordance with Section 277 of the Town Law, and they would become effective after approval by the county board of supervisors. If the area includes a town that has a planning board that has not enacted its own subdivision regulations, the town board may consent by resolution to the county exercise of this power.[16]

Counties may also prepare and adopt official maps. These would show existing and proposed rights-of-way of county roads and drainage systems and the boundaries of sites for public development. This legal tool would help to implement county planning for transportation, flood control, and public facilities. The official map would prevent the encroachment of development on lands proposed for such future county purposes and also would control development along roads and drainage channels or near sites for public development.[17]

STATE LAND DEVELOPMENT CONTROLS

The existing structure of state law that applies to land-use planning and development control is not limited to enabling statutes under which local and regional boards may act. Various state departments and agencies have also been assigned powers and responsibilities for direct intervention in development decisions. No attempt will be made here to inventory all such delegations of state power. The Department of Environmental Conservation and the Adirondack Park Commission are sufficient examples of this aspect of the existing state institutional structure for land-use control that would have to be taken into account if that structure were to be modified to accommodate new responsibilities for coastal zone management.

Department of Environmental Conservation

Article 25 of the Environmental Conservation Law, the Tidal Wetlands Act, gives the Department of Environmental Conservation direct control of tidal wetlands in the state. The act is based on the legislative findings that tidal wetlands are biologically vital and productive areas. They also provide protection from hurricanes and storms by storing flood waters caused by storm surges. Wetlands absorb and convert pollutants and control sedimentation, besides offering recreational, educational, and scientific values. Because vast acreage of tidal wetlands had already been lost to development or despoliation, the legislature found that it is in the interest of the state to preserve as much as possible of the remaining wetlands.

The first step in the implementation of the act is the preparation of an inventory of the tidal wetlands. After public hearing on the proposed delineation of these areas, "the commissioner shall establish by order the final bounds of each such wetland."[18] Meanwhile there is to be a moratorium on alteration of the remaining wetlands. Property owners who feel the moratorium imposes a hardship may apply for a permit to carry out particular activities. The Commissioner of Environmental Conservation may grant a permit after proper hearing. Part 660, Tidal Wetlands–Moratorium Permits, which is part of the Official Compilation of Codes, Rules, and Regulations of the State of New York, gives the procedures to be followed in applying for and granting permits to alter tidal wetlands during the moratorium period.

Once the tidal wetlands inventory is completed, the commissioner and local government officials are to establish a program for wetlands protection. They may enter into cooperative agreements by which state personnel and financial resources may be used "for the purpose of preserving, maintaining and enhancing . . . those tidal wetlands included within the boundaries of such villages, towns, cities and counties."[19]

After the inventory is completed, the commissioner is also to prepare and adopt regulations governing the uses of these lands. He is to determine what uses are compatible with the values of particular wetlands for "marine food production, as a wildlife habitat, as an element of flood and storm control, and as a source of recreation, education and research."[20] The regulations may permit only such compatible uses. A permit from the commissioner, granted after a hearing, will then be required for an alteration of wetlands. This permit will be in addition to, not in lieu of, permits that may be required by the municipality in which the wetland is located. Furthermore, "the applicant shall have the burden of demonstrating that the proposed activity will be in complete accord with the policy and provisions of this act."[21]

The Tidal Wetlands Act provides for enforcement of this permit procedure by declaring that "any person who violates any provision of this act regarding regulated activity shall be guilty of a misdemeanor."[22] Fines up to $2,000 per day of violation may be imposed, and the violator shall also "be liable to the state for the full cost of restoration of the affected tidal wetland to its condition prior to such violation insofar as that is possible."[23]

This is only one example of direct state regulation of land use, in this case by the Department of Environmental Conservation. No state, regional, or local commissions, boards, or governing bodies share this power. The Department of Health and the Department of Environmental Conservation also indirectly influence land development for residential use by their powers to review plans for real estate subdivisions. Part 653 of the Official Compilation of Rules and Regulations of the State of New York, for example, mandates Department of Environmental Conservation review of individual and community water supply and sewerage systems in new subdivisions. This is not land-use regulation by zoning, but it is state participation in subdivision regulation.

An example of the potential impact of state regulation of development through review of sewer and water systems for proposed subdivisions is the Ton-Da-Lay case. In August 1973, the Commissioner of Environmental Conservation, Henry Diamond, rejected the developer's application to construct a major second-home development in the Adirondack Park. The plan called for a community of 20,000 persons on an 18,386 acre tract. Commissioner Diamond based his decision on three factors. Two of them related to the developer's application for a water supply permit and sewerage plans to install individual septic tanks. The third factor, on which the commissioner put most of his emphasis, was protecting the environment for all the people by controlling land use. He said, "What this decision says is simply that we will use all the permit authority vested in the state to try to bring about rational land use rather than sitting idly by drafting lofty land use plans that never really affect what happens on the ground."[24]

Part 615 has added a new kind of development control, which is authorized by a 1972 amendment to Sections 10, 14, and 15 of the Environmental Conservation Law. This regulation calls for environmental impact assessments by applicants for development permits. Such a required assessment "provides a description of the proposed project or development and a detailed analysis of its environmental effects."[25]

The existing legal structure for control of development in the coastal zone and elsewhere by state, regional, county, and local agencies is complex and perhaps uncoordinated. Coastal zone management by the existing structure would be difficult, and changes would be desirable. Nevertheless, existing state laws do provide resource management powers whose present deployment must be taken into account in planning for an alternative statewide management structure.

FOOTNOTES

[1] General City Law, Section 30.

[2] General Municipal Law, Article 12-A, Section 239-a.

[3] General Municipal Law, Article 12-A, Section 236.

[4] Town Law, Section 261 and Village Law, Section 7-700.

[5] Village Law, Section 7-710.

[6] General City Law, Sections 32 and 33; Town Law, Sections 276 and 277; Village Law, Sections 7-728 and 7-730.

[7] General Municipal Law, Section 34; Town Law, Section 278; Village Law, Section 7-732.

[8] General City Law, Section 36. Similar language is included in the Town Law, Sections 279, 280-a and The Village Law, Sections 7-734 and 7-736.

[9] General Municipal Law, Article 12-B, Section 239-d.

[10] *Ibid.*

[11] *Ibid.*

[12] *Ibid.*, Section 239-1.

[13] *Ibid.*, Section 239-m.

[14] *Ibid.*

[15] 31 N.Y. 2d 365, 339 N.Y. Supp. 2d 949.

[16] General Municipal Law, Section 239-d, Paragraph 7.

[17] *Ibid.*, Sections 239-g to 239-k.

[18] Environmental Conservation Law, Article 25, Section 25-0201.

[19] *Ibid.*, Section 25-0301.

[20] *Ibid.*, Section 25-0302.

[21] *Ibid.*, Section 25-0402.

[22] *Ibid.*, Section 25-0501.

[23] *Ibid.*

[24] *New York State Environment*, September 1, 1973, p. 1.

[25] Part 615, NYCRR, Section 615.1 (c).

CHAPTER 10

INSTITUTIONAL ALTERNATIVES

New York State has had a long history of statewide planning. Its activities, status, and place in the state governmental structure have varied over the years. The former Office of Planning Services was located in the Executive Department and was given responsibility by Governor Wilson for coastal zone planning within the program established by the federal Coastal Zone Management Act of 1972. Transition to the Carey administration has located coastal zone planning responsibility in the Department of State and may bring new policies regarding coastal zone affairs.

There is also an established structure of municipal, county, and regional planning organizations in the state. As we have seen, this structure and the powers of the various bodies in its hierarchy have been sanctioned by state law. Governor Carey and the new state legislature may wish to deal with two important issues of state planning: first, the distribution of powers and responsibilities among the state, regional, county, and municipal levels in the planning hierarchy; second, the relationship between planning for coastal zone management and planning for other statewide development concerns that include land use, infrastructure of public facilities, and natural resources.

Full analysis of these issues would require more time and resources than are available for this study of decision-making on the beach erosion problem at Westhampton Beach. Nevertheless, some alternatives for changes or adaptations of the state structure for resources planning and management can be identified and speculated about briefly.

THE ADIRONDACK PARK AGENCY MODEL

The Adirondack Park Agency was created by state law in 1971 and placed in the Executive Department. The agency consists of a board and its staff. The agency board members are: the Commissioner of Environmental Conservation, the Secretary of State, the Commissioner of Commerce, and eight members appointed by the governor. The first mandate of the agency was to prepare an Adirondack Park Land Use and Development Plan. After considerable discussions with local governments and a series of public hearings, the plan was enacted by the state legislature in 1973 as an amendment to the Adirondack Park Agency Act.[1]

The Adirondack Park Plan does not designate land use in detail. It is concerned rather with intensity of use according to the development capability of the land in order to conserve and protect the natural and historic resources of the Adirondack Park. The plan governs the 3.6 million acres of privately owned land (62% of the total) in the Adirondack Park. Six categories of development are designated in the plan:

1. **Hamlet Areas**: existing communities that will serve as service and growth centers. Land-use intensities will be determined by the local governments.
2. **Moderate Intensity Use Areas**: location near hamlet areas, land capability and resource characteristics permit intensity of use up to 500 principal buildings per square mile, an average of one per 1.3 acres.
3. **Low Intensity Use Areas**: physical and biological resources can support some development as indicated by soil characteristics, moderate topographic slopes, and no large areas of critical biological importance. These areas can support up to 200 principal buildings per square mile, an average of one per 3.2 acres.
4. **Rural Use Areas**: areas of shallow soils, steep slopes, significant ecological and scenic values, and generally remote from hamlet areas. Only 75 principal buildings per square mile are permitted, about one per 8.5 acres.
5. **Resource Management Areas**: Management of forest, agricultural, biological, and recreational resources is paramount; and soil characteristics, elevation, presence of scenic rivers, wetlands, and critical biological habitats require limitation of development

to 15 principal buildings per square mile, an average of one per 42.7 acres.

6. **Industrial Use Areas**: Primarily for industrial or mineral extraction enterprises. There are no intensity-of-use restrictions in these areas.

Development in each of these areas would not, of course, take place at uniform densities. The specific characteristics of the property of each developer would suggest how development would be clustered. Local land use regulations would probably also differentiate lower and higher density areas within the overall limitations established in the Adirondack Park Plan. For each category of development, the plan lists acceptable primary and secondary uses.

The Adirondack Park Act requires the agency to encourage and assist the towns and villages in the park to prepare their own land use plans and controls. Few municipalities had them when the agency was established, and there was no regional planning agency comparable to the Nassau-Suffolk Regional Planning Board or others in the state. Local plans that meet certain criteria are accepted by the agency. Local governments with such plans assume a greater share of responsibility for reviewing development proposals.

The Park Plan gives detailed definitions of Class A Regional Projects, for which permits must always be granted by the agency. Municipalities that have agency-approved land use plans and controls may grant permits for Class B Regional Projects, also defined in the plan. In other municipalities, the agency would also control Class B projects. Development projects not in these classes of regional concern would be left to the discretion of local governments.

The Adirondack Park Agency is not the only state agency from which developers in the park must obtain permits. They must also apply to the Department of Environmental Conservation for permits related to water supply and sewerage systems as well as emission of air and water pollutants. Department of Health permits are also required where water supply and sewage disposal are involved as well as for certain recreational activities.

The Adirondack Park Agency model creates a state organization reporting directly to the governor that is empowered by state law to articulate and protect regional and statewide interests in the Adirondack Park. Certain projects of regional significance must obtain agency permits regardless of local land use controls. Outside of hamlet and industrial areas, these could include practically all development, from single homes near rivers and lakes to large-scale subdivisions and commercial or resort projects. Local land-use requirements, where they have been enacted, would also be applied so that developers would have to meet both agency and local regulations. The agency does not have authority, however, to prepare master plans, zoning ordinances, or subdivision regulations and impose them on any municipal jurisdiction in the Adirondack Park. A coastal zone management bill, proposed by one of the Senate sponsors of the Adirondack Park Agency Act, Bernard C. Smith, would grant these powers to a State Commission for Coastal Zone Planning and Management.

THE SMITH BILL

Experience with establishing the Adirondack Park Agency may have been the main inspiration for the coastal zone management bill formulated by Senator Bernard C. Smith, chairman of the Senate Committee on Conservation and Recreation and representative of a Long Island constituency.[2] Suggestions derived from analysis of coastal zone management structure in other states were also incorporated in the bill. Senator Smith intended his proposal to be a study bill that would explore the organizational problems of coastal zone management and provide a focus for discussion and identification of alternatives. The bill will be redrafted after public hearings across the state, but its initial version may provide a starting point for examining changes that may be required in the state laws that delegate land-use planning and control powers.

It has already been noted in the previous chapter that the General City Law, The Town Law, and the Village Law offer a variety of alternatives by which local governments may establish and carry out land-use and development controls under zoning, subdivision regulation, and official mapping procedures. The General Municipal Law provides for county review of local determinations where land parcels lie within 500 feet of jurisdictional boundaries, major roads, drainage channels, and other important features.

The Smith bill would return to the state zoning and land development control powers that had been delegated to local governments for over half a century. These powers, like home rule, are now regarded as permanently vested rights and jealously guarded by cities, towns, and villages. It is assumed in the Smith bill that these powers are not always exercised effectively or competently, and that local bodies tend to ignore vital regional and statewide interests in their planning and development control decisions.

An editorial in *Newsday*, the Long Island newspaper, agrees with State Senator Smith. It quotes him as saying: "Most municipalities in the state just aren't sophisticated enough in their planning to deal with the problem." The editorial goes on to say that "the bill . . . would give the state a veto power over all zoning changes within 1,000 yards of coastlines. The notion is highly controversial because it would impinge on the zoning powers of local governments. We think that impingement is necessary. Here on Long Island, what's left of the shoreline must be protected from the kind of bad decisions that destroyed wetlands and brought the blight of oil terminals and sandpits."[3]

Senator Smith and *Newsday* both assume that members of a state commission and six subordinate regional boards would act more wisely and expeditiously in coastal zone planning and management than either existing state agencies (such as the Department of State and the Department of Environmental Conservation), county and local governments, or existing regional planning bodies (such as the Nassau-Suffolk Regional Planning Board). This point of view is supported by many respected professional planners and attorneys in the field, among them Fred Bosselman and David Callies. They write:

> This country is in the midst of a revolution in the way we regulate the use of our land. It is a peaceful revolution, conducted entirely within the law. . . .
>
> The *ancien regime* being overthrown is the feudal system under which the entire pattern of land development has been controlled

> by thousands of individual local governments, each seeking to maximize its tax base and minimize its social problems, and caring less what happens to all the others.
>
> The tools of the revolution are new laws taking a wide variety of forms but each sharing a common theme—the need to provide some degree of state or regional participation in the major decisions that affect the use of our increasingly limited supply of land.[4]

The Smith bill is long and complex, not easily summarized. Its major concern is the creation of a new institutional structure for coastal zone management: a State Commission for Coastal Zone Planning and Management, six Regional Coastal Division Planning and Management Boards, an Advisory Council to the State Commission, and six Advisory Councils to the Regional Boards. The justification for such a radical structural change is

> that it is apparent that the issues that affect the coastal zone of the state have greatly enlarged and grown in the past few decades and have begun to far outstrip both the interests, abilities, and power of local units to respond to them; that these issues in most cases are of a regional nature although there are others of importance to the entire state and its people; that the local government structures are insufficient in terms of financing and taxing power to fully meet the demands being placed upon them to protect and manage these vital areas for the region and the state; that there is a need to find an effective program of technical and financial aid to deal with problems of primarily local interest, to develop regional bodies with both financial and technical abilities to deal with problems on a regional basis and to provide on a state level for the effective overall management of the state interests in these coastal zones in cooperation with the local and regional bodies.[5]

Responsibility for this mission would be vested primarily in a State Coastal Zone Commission that would be both a planning and operating agency. As it would have power to acquire land and develop and operate public facilities in the coastal zone, it would presumably be able to carry out its plans.

The "coastal planning zone" has been defined in the Smith bill as including all counties that border on salt water, the Hudson River as far as Albany, the Great Lakes, and the St. Lawrence Seaway, as well as the waters out to the international boundary with Canada, boundaries with other states, and to the territorial limits of the United States in the Atlantic Ocean. Within the coastal planning zone, the State Coastal Zone Commission would define a "coastal management zone," within which it would exercise regulatory powers.

The state coastal zone commission members would be appointed by the governor and majority and minority leaders of the two houses of the state legislature. The regional board members would be appointed by county legislative bodies, mayors of major cities, and other political jurisdictions as provided in the bill for each region. Professional qualifications, residence requirements, and terms of office are spelled out in detail. One of its most radical features is that it would give to these appointed officials power to override land and water use regulations enacted by elected local governing bodies. Where local governments had chosen not to enact such laws, the coastal zone bodies would be authorized to impose them according to their own criteria.

The bill contains language to reassure local governments that their powers would not be diminished, but it leaves no doubt that power over land and water use in the coastal zone would rest with the appointed state commission and regional boards. Section 4 of the bill provides that "there shall be partnership of local, regional and state authorities in both the planning and management phases of the coastal zone program" (p. 19). But the local governments are definitely the junior partners. They may adopt local plans, ordinances, rules, or regulations on the use and development of the coastal zone, but only if they "are in compliance with the guidelines and goals of the state plan and program" (p. 20). Local governments may continue to have jurisdiction over building, zoning, and other development permits, but only for "class C" projects, which are defined as "any alteration, repair or improvements not in excess of an increase of twenty-five percent in either size or value of an existing site, structure, use or development within the coastal management zone, *except* for those sites, structures, uses or developments cited as class A and class B projects" (p. 72). The bill takes five pages to list the class A and class B projects that are determined by the bill drafters to be of statewide or regional significance.

The Smith bill also tries to reassure local governments that their power and authority will not be diminished by the creation of a state coastal zone commission "except as expressly provided in this act." Local governments would not be precluded from adopting or enforcing plans, ordinances, rules and regulations "if such plan, ordinance, rule, or regulation is in compliance with the guidelines and goals of the state plan and program." Any land and water use and development ordinance enacted by a local government would be valid and enforceable "unless the state commission or a regional board has acted contrarywise." Private developers would be bound in land development or subdivision projects by local ordinances "unless the state commission has acted contrarywise."[6]

What the bill says is that local governments will retain their planning and land regulation powers, if they do what the state coastal zone commission tells them to do. At public hearings on the bill chaired by Senator Smith, however, representatives of local governments and citizen groups universally opposed the bill because it infringed on local home rule.

The Smith bill does not simply limit local planning and zoning to enforceable regional and state guidelines. It also gives the state commission and the regional boards power to intervene in the local planning and zoning process. Section 16 of the bill provides that "within one hundred eighty days of enactment of this act, each local government shall submit copies of its local land and water use plan and ordinances . . . to the regional board having jurisdiction over its territory. When a local government fails to submit a local plan or ordinances by the specified date, *the regional board shall prepare and issue a plan for that area*" (p. 44; emphasis added).

The regional boards would notify local governments when they disapprove local plans and land development controls. The local government would have three months to comply with regional board demands for changes, but if within that time the local government has failed to make such corrections, "*the regional board shall issue planning and zoning regulations* which shall govern . . . land and water use and development for that portion of the coastal zone" (p. 47; emphasis added).

The regional boards are not the final authority, however. This is in the hands of the state commission. If it finds that a regional board is not conforming to state commission planning guidelines or is not producing its regional coastal zone plan on schedule, then "the state commission will assume the authority and responsibility for the planning of the coastal division involved."[7]

These proposed state delegations of powers to regional boards and a state commission to override local planning and zoning determinations may or may not be a good idea. That is not an issue to be decided here. But the Smith bill, or its successor versions, should make clear how they change the effect of the General City Law, the Town Law, the Village Law, and the General Municipal Law. Amendments to these laws would appear to be required to give review authority and even direct planning and zoning powers to the proposed regional and state coastal zone bodies. Modifications of other state laws may also be required to create the institutional structure proposed by the Smith bill. This is because it attempts to transfer planning and zoning powers delegated by state law from elected governing bodies to appointed regional and state bodies. If this is what the drafters of the Smith bill want, then it appears they should follow the example of the Suffolk County Charter (Section 1330) and make explicit their intention to limit municipal planning and zoning authority. It would appear that an electoral mandate might be required, or at least action by the state legislature.

STRENGTHENING EXISTING PLANNING INSTITUTIONS

Regional and statewide concerns in land-use planning, which includes land management in the coastal zone, have not been articulated clearly or forcefully enough to influence state policy. There are many statements about such policy in various state laws, but with the exception of the Tidal Wetlands Act not much has been done. Land-use planning and development control powers have been delegated to local governments. Metropolitan, regional, and state planning agencies exist, but they are mainly concerned with research; and their planning proposals are advisory rather than controlling in local decision-making.

Nevertheless, state law does provide for a hierarchy of municipal, county, regional, and state planning organizations that could provide complete coverage of the territorial and functional aspects of planning. What is missing in the present institutional structure is mandatory review procedures that give higher levels in the planning hierarchy real power to articulate development guidelines and insist that local planning and land development controls take them into account. There is one exception to this weakness in the institutional structure for planning, the Suffolk County Planning Commission.

Development of County Powers

The powers of the Suffolk County Planning Commission are not broad enough to bring countywide policies to bear on all aspects of local planning, but they do indicate a direction for further exploration. It has already been noted in Chapter 9 that the Suffolk County Charter has limited local zoning powers by granting the County Planning Commission veto power over local zoning determinations where there are municipal land-use conflicts in the strip of land 500 feet wide along each side of municipal boundaries.

In this respect the Suffolk County Charter goes beyond the General Municipal Law and supersedes it, according to the Smithtown case, decided in the Court of Appeals. The charter goes beyond the state law in another provision that extends County Planning Commission mandatory review of local zoning, subdivision control, variances, and special permit determinations to the strip of coastline 500 feet from the marine edge.[8] This device provides for comprehensive review of development proposals along Suffolk County's 600-mile shoreline. The strong, technically competent staff of the Suffolk County Planning Commission is in a position to evaluate local determinations from a countywide perspective, and its recommendations must be overturned by "vote of a majority plus one of all the members of the referring body in a resolution that sets forth its reasons for not adopting the planning commission's recommendations."[9] It is not likely that the coastal zone can be limited to a strip 500 feet wide. But modification of state and county law to amplify this territorial definition and grant the county the same veto power over local zoning decisions in this area that it has along the boundaries of municipalities would go a long way toward strengthening county participation in an intergovernmental structure for coastal zone planning and management.

There may be some disagreement about the suitability of the county as the appropriate level of government to review local development decisions, especially for coastal zone management. It may be argued that all county governments are not as sophisticated in structure, expertise, and leadership as Suffolk County. This may be true, but county government is blossoming nationwide under pressure from the demands for authority and services not available to individual municipalities. What is suggested here is that the resources that would be devoted to creating a new institutional structure as suggested by the Smith bill might better go to strengthening county government in the state. Basic to such development, however, would be the willingness of county voters and political leaders to reform their governmental structure and grant the necessary powers and resources by approving county charter referendums.

Existing Regional Planning Commissions

Regional planning bodies cover the entire marine and freshwater coastal areas of New York State. They are multicounty agencies established under the authority of the General Municipal Law. As they are not regional governments with elected legislatures and executives, regional plans prepared by them have no legal effect unless appropriate segments are adopted by county governments. For the same reason, regional planning organizations have no legal authority to implement plans by means of public works construction or regulating private development. Nevertheless, they have potential for exerting considerable influence on county and municipal development decisions.

Given the existence of such regional planning bodies, the question arises why another layer of coastal zone boards, as proposed in the Smith bill, would be required. If the existing agencies should have additional powers and resources to carry out the functions visualized by that bill, why not amend state law to strengthen the present structure? Building up these organizations would also assure the integration of coastal zone planning with other aspects of regional planning.

The Smith bill recognizes the need for such integration, but apparently does not visualize the limitations of its proposed regional boards to do this. The bill provides that regional coastal zone plans shall include: "recommended plans for regional transportation systems, recommended plans for a regional solid and liquid waste disposal system; recommended plans for regional water distribution systems and recommended plans for regional energy distribution systems."[10] The drafters of the Smith bill appear not to realize that a strip of contiguous counties along the shore is not an appropriate "region" for planning transportation, waste disposal, water, or energy distribution systems. Long Island is only one county wide, but the Great Lakes and Hudson River segments of the coastal zone are integral parts of extensive regions delimited in various ways by drainage basins, transportation and communications networks, economic structure, and other factors. However "regional planning" may be defined, the coastal strip is not a likely framework for it. A more effective approach, it appears, would give special emphasis to coastal zone problems within the context of planning for more comprehensive regions.

It is this point that leads to questions about whether coastal zone management needs a new set of institutions added to the already complex structure of existing governmental agencies and jurisdictions. Policy-making, planning, and action are already plagued by a maze of intergovernmental and interorganizational negotiations. Further complicating the institutional structure would appear to proliferate negotiations and compromises, encourage bargaining away of vital environmental values, and make decision-making more tortuous.

If the time has come that the state's coastal resources are so threatened by population and economic growth pressures that the legislature must be directed, as in the Smith bill, to enact a coastal zone plan, why does this not also apply to resources in other areas of the state? If the time has come for the state government to take state and regional planning seriously, should it not undertake the process comprehensively rather than segmentally? If the state legislative and executive arms are indeed willing to make development policies and programs explicit, experience should perhaps be given time to indicate whether such policies can be passed down to existing state and regional bodies for implementation or whether new institutions must be created. For example, would a new Long Island regional coastal zone management board, as proposed in the Smith bill, be able to do a better job than the existing Nassau-Suffolk Regional Planning Board? Why not grant the necessary powers and financial resources to the existing organization rather than to a new duplicate body that would take years to gain the experience and knowledge already available in the old one?

Coastal Zone Planning at the State Level

With the change from the Wilson to Carey administrations, state planning in New York is likely to undergo another of its periodic reorganizations. Although its future shape cannot be foreseen, it may be assumed that state planning will not be abandoned. One hopes it will be strengthened to provide useful information inputs for state policy formation on resources management, including coastal zone resources, as well as to formulate a statewide development framework for coordinating municipal, county, regional, and state planning. The former Office of Planning Services, located in the Executive Department, was designated by Governor Wilson to be the state agency to prepare the New York State coastal zone management program in compliance with the federal Coastal Zone Management Act of 1972. In the Carey administration, the Secretary of State has assigned the Division of State Planning the critical task of articulating statewide interests in the coastal zone and integrating local and regional planning efforts into a state program.

It has been suggested that regional planning involves coordinating concern for coastal zone resources with the development and conservation of all regional resources. State planning should have the same comprehensive concern for state resources. A state coastal zone commission, cut off from the central state planning agency as recommended in the Smith bill, would complicate the institutional structure and likely cause the proliferation of bureaucratic and political struggles for power, jobs, and money. This would hardly advance state interests in resources management.

A tentative suggestion that might be put forward for redefining the role of a state planning agency would posit the creation of a state resources planning board. The board itself would represent political, bureaucratic, and citizen interests and serve primarily as an advisory body for the governor. The organization staff would be responsible directly to the governor but also provide technical support for the board. An oversimplified communications network is illustrated in Figure 19 to indicate the board's structural relationship within the state governmental hierarchy. The board is given a central position in the diagram because it is the focus of attention of this discussion, not because it is proposed to overshadow other organizations.

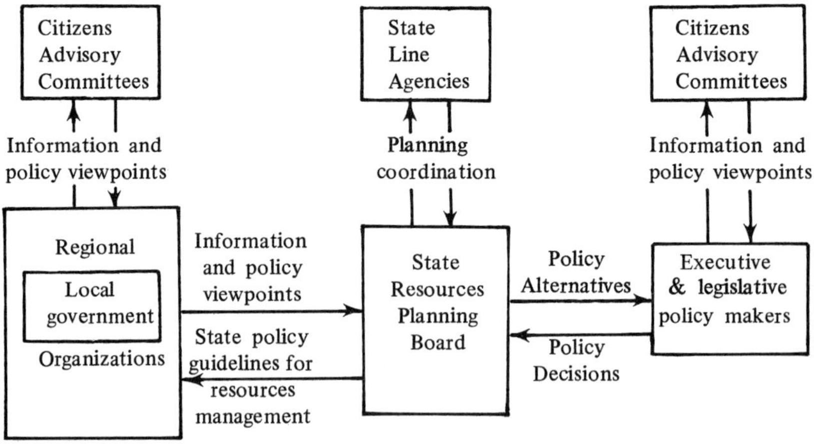

Figure 19. State Resources Planning Board communications network.

Effective communications would, of course, depend on the willingness of state and local agencies to permit a state resources planning board to function. Bureaucratic rivalries could put it out of business in short order. Effective input into governmental decision-making would depend on the quality and relevance of the information and policy analyses produced by the board. Experience with planning organizations at all levels of government has revealed the fundamental difference in the approach that politicians and technicians take toward the articulation and analysis of policy issues. The technical approach is to define the issues in terms of their relation to particular systems and to identify changes required in the systems to resolve the issues and deal with the problems involved. The politician is concerned with gaining and conserving power and the support of identified constituencies. His decisions are generally based on concern for political costs and benefits for himself rather than primarily on improvements in the systems that the technicians have modelled.

Anthony Downs has pointed out another structural problem with which a state agency must contend when its mission, among others, is coordination of other agencies:

> In my book *Inside Bureaucracy*, I call it the "Law of Imperfect Control." It states that within any large organization . . . it is impossible to achieve perfect coordination, and the bigger the organization, the less adequate the coordination. In really large organizations . . . this law clearly implies that it is impossible even to achieve adequate coordination among the key actors in advance of their taking action. There are too many actors working on too many complex problems that are related to each other in nonobvious ways and pursuing too many different goals, for adequate coordination to emerge in advance. That is why action is disjointed.[11]

Nevertheless, large and complex organizations, such as state government, will continue to exist. Planning will necessarily become an increasingly important function in such organizations, whether or not their managers realize it. A central planning organization in state government can perform effectively if it is successful in managing the functions and conflicts inherent in complex bureaucratic-political structures.

PLANNING AND IMPLEMENTATION

American governmental institutions generally assign planning and implementation to different organizations. At the local level, for example, planning boards and commissions can only propose directions for future community development. Public works to implement plans are decided in the city council or town board and carried out by line agencies. The local planning agency administers land use regulations, but only after they are enacted by the legislative body. Metropolitan and regional planning agencies have neither implementation nor direct land-use control powers. They may have advisory review jurisdiction over local zoning decisions, and they have advisory review of projects that depend on federal financial assistance. At the state level, the Division of State Planning in the Department of State also has no implementation powers.

The exception to this split in responsibility for planning and implementation is the public authority. Once established, it prepares its own plans, finances its projects by selling revenue bonds or even levying authorized taxes, and proceeds with construction, directly or through subcontractors. Examples of such authorities in New York State are the Port Authority of New York and New Jersey and the Urban Development Corporation. The latter organization goes beyond the single-purpose objectives of most public authorities, such as the New York State Thruway Authority, which was established only to build that particular highway. The Urban Development Corporation is not limited to a single function, such as housing. It is sponsoring the construction of entire new communities and experiments with new construction technology.

Although it is beyond the scope of this report and the competence of its author to propose definitive solutions to the problem of building coastal zone management institutions, it appears that this process would need to move in three directions at the same time.

1. To avoid the extremes of freezing private development investment or rewarding exploitive speculation, develop institutions to define the carrying capacity of fragile environments and acquire for the public either ownership or development rights

to lands designated for conservation or preservation. Tax abatement is another device to avoid private land confiscation by zoning and other controls.

2. To protect statewide interests in the coastal zone, institutionalize the review of development decisions at the local, county, and regional levels to assure conformity with state coastal zone policies and development guidelines.

3. To carry out coastal management projects for pollution control, shore stabilization, recreational use, etc., establish public authorities to: (a) identify development problems, (b) combine local, state, federal, and private resources, (c) participate in design, financing, and construction.

One issue to be resolved in institution-building for coastal zone management is whether or not to combine planning and implementation/enforcement authority in the management organizations. There are advocates of retaining the present institutional system in which planning agencies have no other functions and are presumably not distracted by management responsibility. Dr. Lee Koppelman, Director of the Nassau-Suffolk Regional Planning Board and also of the Suffolk County Planning Commission, is one such advocate. Although he does not favor acquisition of management responsibility by his own regional planning agencies, he does not have confidence in state management capabilities. Dr. Koppelman's view stems from his observation of the weakness of state enforcement of existing environmental laws and regulations. For example, he believes that pollution in the south shore bays could be eliminated if there were adequate state enforcement of environmental laws and regulations already on the books.[12] His preference is for county responsibility for management, particularly in the area of development regulation. His position was stated in a newspaper article that reported progress on a coastal zone guidelines study. He was quoted as saying:

> In this study we are required to include alternatives by which coastal zone plans can be implemented, and we may recommend some changes in where the final decision-making power should lie. But we do not want to take zoning power away from municipalities. . . . I personally believe that municipalities

> should make the decisions for their coastal zones in line with regional guidelines, but that the county should have the right to overrule them if what they do is out of line with the overall planning for Long Island.[13]

This would seem to imply extension of the Suffolk County Charter review power granted the County Planning Commission over local zoning decisions along municipal boundaries and along the marine shoreline. These powers were granted specifically to the Planning Commission, which thus combines planning and implementation authority for certain countywide concerns. Dr. Koppelman apparently does not want to see this combination of planning and implementation/enforcement power vested either in the state or in his own Nassau-Suffolk Regional Planning Board. The news article cited above concludes with this paragraph:

> A controversial study bill sponsored by State Sen. Bernard Smith (R-Northport) would put a new state agency atop a pyramid of governments participating in coastal control. Koppelman prefers to keep the final say out of state hands by giving it to the counties. Most towns and villages want no interference with their existing "home rule" powers, but seem receptive to having the regional board produce guidelines.[14]

Plan implementation goes beyond higher-level review and control of local land use control decisions. It includes the actual construction of the public works proposed by the plan. Separation of planning and the construction phase of implementation could be potentially dangerous. A "general plan" is essentially a concept of an idealized future and is not necessarily bound by practical constraints. Such a plan may raise expectations in the minds of private developers and public agencies that will not be fulfilled. Publication of a state coastal zone plan, for example, could trigger local action to either limit or promote development in the expectation that certain coastal zone projects would be carried out. But they may not, for the plan would not necessarily be implemented.

The Corps of Engineers plan for constructing 50 groins on the barrier islands and beaches from Jones Inlet to Montauk Point was such a general plan. Development pressures were aroused by the prospect of this program of shore stabilization, which appeared to assure the safety of homes and commercial establishments on the barrier beaches. Nevertheless, the project is making very slow progress. Reaches of the barrier islands still not protected by the long-range programs of the Corps of Engineers suffer heavy financial damages from beach erosion and storm surge flooding. Had not the "general plan" of the Corps offered the prospect of shore stabilization, stronger development restrictions might have been imposed by local governments; and private investors might have been reluctant to invest in real estate threatened with obliteration. The point is not that the Corps of Engineers should have prevented publication of its Jones Inlet to Montauk Point survey and plan reports. This example is simply cited to show what may happen when implementation is not built into the total planning-management process.[15]

The relation between planning and implementation, for the coastal zone and elsewhere, points up some practical and philosophical problems in the development process that are still unresolved. They may be expressed in the two related polar antinomies of centralization *vs.* decentralization of decision-making, and comprehensive *vs.* incremental planning. The decentralized, incremental decision model may be exemplified by uncontrolled economic and political markets for goods, services, influence, and power. In such a mythical system numerous individuals and organizations bargain and exchange values according to their own perceptions of costs and benefits. The result is supposed to be economic efficiency and maximum satisfaction for society as a whole. The model of centralized comprehensive planning at the other extreme is authoritarian government.

The actual situation is somewhere between these extremes, and theoretically perfect markets are replaced by some degree of corporate and governmental control. The problem, therefore, is how to maintain economic and political freedom while at the same time preventing excess concentration of economic and political power in either private or public hands. A parallel political problem is how to attain some equitable distribution of income and influence.

In coastal zone planning, federal law and state objectives mandate local government participation and the protection of individual rights, but increasing pressure on coastal resources requires high-level determination of development policy. We return, then, to the problem of how to design an institutional structure that can accommodate the formation of broad, coherent state policy on the conservation and development of coastal resources and still encourage local self-determination within that framework. The crucial problem is how to determine the direction and constraints on development within the limits of state coastal zone resources in a context of equity and social justice.

The attractive suggestion that planning and implementation functions be combined in the same organizations would foster the incremental paradigm, for each agency would pursue its own limited objectives. So would insisting on the preservation of local prerogatives and veto power over higher-level decisions. Conflicts of parochial jurisdictional interest and agency interest would have to be discerned and reconciled so that coastal management might have coherence and the probability of operational success. Perhaps the most arduous tasks in policy-making are determining the big picture of development for the whole society and then exercising the moral and political fortitude to see that those policies are carried out.

FOOTNOTES

[1] Information about the Adirondack Park Agency was obtained from the Adirondack Park Land Use and Development Plan and Recommendations for Implementation, March 1973, Ray Brook, New York.

[2] S.9586, introduced March 18, 1974.

[3] *Newsday*, January 2, 1974.

[4] Bosselman, Fred, and David Callies. *The Quiet Revolution in Land Use Control* (1971).

[5] S.9586, Section 2, Legislative Findings, p. 5.

[6] S.9586, Section 4, pp. 19-20.

[7] S.9586, Section 16, p. 50.

[8] Suffolk County Charter, Sections 1131, 1323, 1332, and 1333, paragraph (6) in each section.

[9] Suffolk County Charter, Section 1325.

[10] S.9586, Section 15, p. 41.

[11] Downs, Anthony. "What This Profession Really Needs is 5 Cents Worth of Irrationality," *Planning*, July 24, 1974.

[12] Interview June 18, 1974 at Suffolk County Center, Hauppauge, New York.

[13] Morris, Tom. "Coastal Zone Guidelines a Step Closer," *Newsday*, November 12, 1974.

[14] *Ibid.*

[15] Interview with Beecher Greenman, consulting engineer, June 25, 1974.

CHAPTER 11

CONCLUSIONS

This investigation has been concerned with the general problem of structuring governmental institutions and decision-making for coastal zone management. It was not based on any *a priori* hypothesis about how this problem should be solved, nor has it sought conceptual or empirical support for such a hypothesis. Instead, a particular problem of coastal zone management was selected for study to identify existing intergovernmental decision structures for dealing with it. Analysis of this problem may lead to evaluation of the adequacy of existing institutional arrangements and identification of possible directions for change, if change is required.

The first step in this study was the attempt to describe the nature of the physical coastal management problem itself—in this case the control of erosion at Westhampton Beach. It became apparent from examination of some of the literature and from discussion with scientists and engineers who had studied the problem that there is considerable uncertainty as to its causes and disagreement about what to do about it. The next step was to identify the governmental organizations responsible for decision-making, their evaluation of the problem, and their position with regard to possible action alternatives.

EROSION AT WESTHAMPTON BEACH

Ocean beaches are dynamic geomorphic structures. They erode or build up under the influence of seasonal climatic changes, the availability of sand for nourishment, and the traumatic effects of severe storms. Although the general pattern of beach dynamics is well known, there are still many uncertainties about the laws by which the variables of wind, wave movement, and sand particle characteristics relate to each other to produce particular effects on particular beaches.

It has been noted that the prevailing littoral drift along the ocean beaches of Long Island is from east to west, that sand is carried from the eroding Montauk bluffs to the barrier islands, and that much of the sand in the littoral current is trapped by Shinnecock Inlet and deposited in the bay inside the inlet. Loss of sand from the shore current deprives the westerly beaches and dunes of material for maintaining stable configurations and makes them vulnerable to erosion by severe wave and wind action. It has also been noted that man's activities affect the beach and dune structure and thereby intensify damages on the barrier islands and bay shores from hurricanes and northeasters. Attempts to stabilize the beaches by building groins may even aggravate the erosion of beaches to the west of these structures. Nevertheless, the specific causes of beach erosion on Westhampton Beach are imperfectly understood and, consequently, so are the effects of alternative action proposals for dealing with it.

INFLUENCE OF TECHNICAL ANALYSIS ON POLICY-MAKING

One of the objectives of this investigation has been to discern from technical studies the nature of the erosion problem at Westhampton Beach and to see how this information has been used in deciding what to do about it. Several scientific studies have been cited as well as the engineering investigations of the U.S. Army Corps of Engineers. The current state of knowledge about beach dynamics and its application to dealing with problems of erosion have been summarized very well in two of the publications cited, *Coastal Geomorphology* and *Coast Stabilization and Protection on Long Island.*[1]

Nevertheless, in spite of strongly held theories about how and why barrier islands change over time, and how these theories apply to Long Island ocean beaches, there appears still to be considerable uncertainty about how sand grains are moved by wind and water at particular times and places. There are also fervent advocates of divergent approaches to dealing with the problems of beach erosion. One school advocates engineering works to reconstruct beaches and dunes, with sand dredged from ocean and bays, and stabilization of the beach fill with stone groins. Others say let nature take its course and let storm surges wash over the barrier islands. Still others take a middle ground in favor of helping nature along by replenishing sand on the beaches and in the littoral current and by devising some kind of mechanical means to bypass sand across the inlets between barrier islands. Unfortunately, a successful sand bypassing system has yet to be designed; and thousands of tons of sand deposited for beach nourishment one year may be washed away the next.

The federal government, through the Corps of Engineers, has adopted the engineering works strategy and advocates completing the groin project at Westhampton Beach. The state apparently goes along, and the Town of Southampton agrees with this approach. The County Public Works Commissioner supports the completion of the groin field with shorter and less expensive structures, with beaches filled to the ends of the groins. The Suffolk County Legislature, however, has changed its mind. Until the November 1974 elections, most legislators favored completing the Corps of Engineers project. The new legislature has refused to go along.

Neither state nor county agencies have undertaken their own scientific or engineering investigations. The Nassau-Suffolk Regional Planning Board has conducted no original research in this field either, and it has not been asked to make a specific technical judgment on the Westhampton Beach erosion problem. Its publications, however, appear suspicious of the engineering works approach and say that groins should not be built unless there is good technical justification for them. County Executive Klein has interpreted this position to mean that the Nassau-Suffolk Regional Planning Board advocates the "let nature take its course" theory and opposes all groin construction. True or not, this is the position firmly adopted by Mr. Klein.

DECISION-MAKING IN COASTAL ZONE MANAGEMENT

Another objective of this study has been to consider the problem of institutional arrangements for intergovernmental decision-making in coastal zone matters. It had been assumed that if technical knowledge about coastal problems was quite reliable, then the institutional problem would be concerned only with assuring that the information was made available to all the decision-makers involved and that effective communications between them would lead to technically rational and politically authoritative decisions.

Unfortunately, where there is uncertainty about the nature of the questions that must be decided and about the consequences of alternative courses of action, improving institutions cannot contribute much to improving the decisions. If reliable knowledge is not available to guide decisions, then improving the decision-making structure is of little avail for problem solving. Where there is technical uncertainty, institutional structure is not the obstacle to dealing with coastal problems. Improving the structure will not improve decision-making where the constraints on problem-solving are in the nature of the problems themselves, not in the institutional arrangements or decision process.

In spite of this difficulty, devising an effective structure for intergovernmental decision-making in the coastal zone has drawn the attention and efforts of political leaders and administrators. First, it is required by the national Coastal Zone Management Act of 1972. Second, not all coastal zone problems are technically intransigent, and facilitative institutional structures may afford the means of helping to cope with uncertainty. Third, technical knowledge about coastal problems will improve over time, and the best management institutions that can be devised should be available to take advantage of it.

It is apparent that many governmental jurisdictions are involved in coastal zone management. It is the essence of the federal system that each jurisdiction, and the constituencies represented in it, have some voice in decisions that affect their own interests and welfare. Nevertheless, differences of value and opinion within and among jurisdictions will lead to inevitable conflicts. Management decisions must, therefore, be the outcome of political and financial negotiations that seek accommodation of the divergent interests in the coastal zone.

The system of negotiations relating to the beach stabilization and hurricane protection problem at Westhampton Beach was deliberately designed to be decentralized. It requires decision-making consensus and financial cooperation for projects to be planned and carried to completion. In this particular case, federal, state, and county jurisdictions are involved, and each one has veto power.

A federal system is one that governs by negotiation and accommodation among the various levels of public jurisdiction. Serious problems do arise, however, when the public interest in the more inclusive jurisdictions must override the interests of the smaller, more exclusive ones. The political issue, then, is to determine which social problems merit such centralization of decision-making power and how the general public interest may be defined out of the welter of private interests.

Until recently, shore erosion and other problems in the coastal zone were not singled out as a special category for political and technical decision-making. The Coastal Zone Management Act of 1972, however, has given these problems new status as a separate focus of national concern. It offers inducements to states to create special planning and management arrangements for the coastal zone, and a new political and bureaucratic apparatus is developing to control federal-state relations in this area. State response to these national inducements, in turn, requires the establishment of a separately identified state-local planning and management apparatus for the coastal zone. Action has already been initiated by New York State as the result of its award by the National Oceanic and Atmospheric Administration of funds to support the planning phase. Legislators have also prepared bills to establish the management structure. The major question in these efforts is whether coastal zone affairs will be managed by existing political and bureaucratic institutions or whether new ones will be needed.

A new institutional structure for state-local relations in coastal zone management could result in greater centralization of decision-making based on the rationale that increasing demands on coastal resources require that state-wide interests override local interests. One way to bring about such centralization is by constitutional and statutory changes that would retrieve to the state powers that had previously been delegated to local jurisdictions. Another would be for the state to assume full responsibility for financing public works for coastal zone management and for assuming the full nonfederal share of the costs of cooperative projects.

It is not likely, however, that local jurisdictions would favor recentralization of planning and management authority in the state, even if this meant relief from some financial burdens. Suffolk County Legislature Representative H. Beecher Halsey, Jr. notes that there is strong home-rule sentiment in the area. People in the towns and villages of Suffolk County would prefer to retain such powers as they now have to either favor or oppose state proposals for public works or management systems in the coastal zone.[2]

THE INFLUENCE OF SOCIAL VALUES

Centralization or decentralization of decision-making is perhaps the overriding issue for institutional change in coastal zone management. It is paraphrased by such expressions as conflict between "parochial *vs.* regional interests" or "local *vs.* cosmopolitan concerns." It is the natural tendency of many political leaders, administrators, and concerned citizens to put their trust in laws, regulations, organizations, and enforcement procedures to solve problems of management of resources for which government has taken responsibility. Nevertheless, there is ample evidence that formalized structures and procedures may be corrupted by those who operate them, or they may fail to attain their objectives because of general public disregard for them.

The most serious questions for institutional change in coastal zone management are: Who shall make policy and exercise management authority—regional or local organizations? What shall be the content of policy and management criteria—advancement of individual and corporate economic interest or protection of the communal stake in environmental values? Finding answers to these questions will involve dealing with the problem of determining trade-offs between community needs for population and economic growth and conservation of environmental resources. How these questions are answered will depend on how much public determination may be mustered to substitute values that favor community and life-supporting environmental concerns instead of the political focus on preserving the rights and privileges of those who seek only private short-run gains from exploiting the environment.

Management of the coastal zone in New York State will involve determination of the geophysical and ecological constraints that ought to limit population and economic growth in particular areas during particular time periods. As market demands, the state of technology, the depletion of nonrenewable resources, and other parameters change, estimates of the population and productive carrying capacity of the environment may also change. To keep options open for future growth and to care for the welfare of future generations, we apparently should avoid actions now that would cause irreversible environmental damage and future deprivation of environmental resources. Technical knowledge for determining the current carrying capacity of particular coastal environments is admittedly limited. Technical uncertainties must, therefore, be compensated by social and political value judgments in making development policies. In this process one can only hope that public constituencies and political leaders will be sensitive to environmental values and prudent in the exploitation of nonrenewable resources.

Finding answers to the questions about institutional change for coastal zone management may also involve reexamination of the constitutional distribution of state and local powers. This would provide the basis for deciding on recentralization of power, if necessary, to determine environmental policy and to supervise and coordinate local action to advance identified regional environmental interests and values. Effective management of the coastal zone may very likely require some strengthening of constitutionally sanctioned state control. But this may have negative consequences if provision is not made for local participation in policy-making. Perhaps more important for successful centralization of coastal zone management authority would be avoiding corruption of the process by struggles for political advantage in the legislature, empire-building in executive departments, or political hesitancy to limit opportunities for private exploitation of coastal zone resources.

These are only generalizations. There are no formulas for institutional change that will guarantee successful and equitable planning and management for coastal zone resources. Institutions, formal and informal, will evolve out of innovative proposals, debate on them, legislation, and practical experience. Formal organizations are only part of the management structure. Equally important are the informal governmental relations that develop out of bargaining among the interests involved and out of the values of public concern and good will that may inform these negotiations. Trust in the efficacy of formal organization and management procedures may have to be complemented by faith in human social instincts.

CURRENT STATUS OF THE PROJECT

The final decision has not yet been made about whether or not to build six more groins and reconstruct the beach and dune at the western end of Westhampton Beach. The Barrier Beach Association in the Town of Southampton has filed suit against both the federal government and Suffolk County in federal court.[3] The plaintiffs requested a court order to compel the United States and Suffolk County to proceed with construction of the project and also demanded $7 million as compensation for property damages caused by failure to provide the shore erosion control and hurricane protection works originally planned.

The first claim by the plaintiffs, for a court order to compel construction, was apparently settled without trial by agreement between the Barrier Beach Association and the United States. This took the form of a stipulation filed August 13, 1974, in the U.S. District Court in the Eastern District of New York. This stipulation, or agreement, has several remarkable features.

It noted that funds had been or were shortly expected to be approved for project construction; and that because the shore was vulnerable to storm attack, there was imminent danger of destruction of life and property. It also noted that "completion of the preparation of the environmental impact statement for approval by the Council for Environmental Quality may take considerable time, as much as one year." In view of these findings of imminent danger, the parties to the stipulation agreed that it would be in the public interest for the Corps of Engineers to initiate construction as soon as possible, even *before* final approval of an environmental impact statement by the Council on Environmental Quality.

Suffolk County refused to sign the agreement. In a letter to U.S. District Judge John R. Bartels, attorneys for Suffolk County pointed out that the county had not entered into the stipulation, and that it did not agree that there was imminent danger to the barrier beach.[4] Suffolk County Executive Klein was reported still to threaten veto of any County Legislature resolution approving county participation in the project, and a *Newsday* editorial characterized it as another unwanted public improvement that was part of the federal pork barrel.[5]

In view of the stipulation between the Barrier Beach Association and the Corps of Engineers, the Corps scheduled a hearing for December 9, 1974. The notice of hearing did not mention groin construction at all. It was to be concerned only with the proposal to dredge material from the bottom of Moriches Inlet and Moriches Bay for deposition on Westhampton Beach. Proponents and opponents of the project were drawing up their positions on whether there was imminent danger to the barrier beach and whether or not groins were helpful, when nature took a hand in the controversy. According to a December 3 *Newsday* report,

> At 10:30 a.m. yesterday, the swirling white foam was still rushing under houses and over dunes, spilling debris and sand onto the street, then continuing its relentless path into the bay. A crowd of county officials and property owners gathered to survey the damage wrought by the northeaster that howled all night, stirring the ocean to a destructive fury. For many it was a familiar scene—the ocean has broken through the barrier beach here periodically since the hurricane of 1938. But the significance

> of the timing was not missed. The storm—and the damage—came exactly one week before a public hearing on a controversial Army Corps of Engineers proposal to build six groins at the beach, groins that property owners maintain will prevent such breakthroughs and erosion.
>
> "That's Klein's inlet. That's the way he wants it," said Fred LaVigna.[6]

At the public hearing on December 9, 1974, the barrier beach residents were out in full force. They pointed out what they thought was the obvious lesson of the storm of the previous week: the washover had occurred where the Corps of Engineers proposed to complete the groin field; there was no damage where the beaches were stabilized by the groins already in place. Nevertheless, the opponents of the project maintained their positions. County Executive Klein said he would veto any County Legislature attempt to approve participation in the project. Other opponents argued that it would be illegal and inappropriate to go ahead with the project until an environmental impact statement had been prepared and analyzed, even if this would take at least 18 months, the time required as estimated by the Corps of Engineers. Both sides rehashed all the arguments that have been discussed in previous chapters of this report. Positions appeared to have hardened, and ideology, emotions, and vested interests were appealed to as often as scientific evidence.

A spokesman for the Office of Louis Lefkowitz, State Attorney General, questioned the usefulness of the proposed groins and suggested that the Corps of Engineers would be on doubtful legal ground if it proceeded with the project. He questioned the legality of a federal court attempt to override the National Environmental Protection Act by waiving the environmental impact statement requirements. He also doubted the legal validity of the stipulation filed in the federal court, because it was agreed to only by the Corps of Engineers and the Barrier Beach Association. Suffolk County had not signed it.

On December 20, 1974, the Suffolk County Legislature decided the issue for the time being. It voted down a proposed resolution to approve county participation in the project and appropriate $445,000 as the county financial contribution.[7]

According to the local cooperation agreement required by the Corps of Engineers, the county would be expected to provide lands and rights-of-way required for construction and also to hold the federal government harmless for any private property damage that might result from the project (see Chapter 7). Presumably the project could go ahead if the State of New York would assume these responsibilities and also provide the 9% share of the capital cost that had been assigned to Suffolk County. The state is unwilling to do this. A representative of the State Department of Environmental Conservation had said at the public hearing on December 9 that the project must be acceptable to the local people. The state would therefore not state its position until the Suffolk County Legislature had made its determination. Presumably the state will not act now that the county has refused to cooperate.

The Westhampton Beach project is apparently dead. So, it would appear, is the rest of the Fire Island Inlet to Montauk Point shore stabilization and hurricane protection project approved by Congress in 1960. Nevertheless, the Barrier Beach Association is pursuing its suit. It is now aiming at a federal court order to compel the state and county governments to participate in the project. This is a possible, but doubtful, outcome.[8]

FOOTNOTES

[1] Coates, Donald R., Ed. *Coastal Geomorphology*, and F. L. Bartholomew and W. V. McGuinnes, Jr., *Coast Stabilization and Protection on Long Island.*

[2] Interview June 12, 1974 at Suffolk County Center, Riverhead, New York.

[3] Thomas O'Grady and Dorothy Patton, individually and on behalf of all others similarly situated as members of the Barrier Beach Association in Southampton Township, Suffolk County, New York against United States of America and Suffolk County of New York, 73 Civ. 1182.

[4] Letter to Judge Bartels from Ronald L. Ginns, dated August 9, 1974.

[5] *Newsday*, October 8, 1974.

[6] Snider, Jane. "Angry Sea, Angry Residents," *Newsday*, December 3, 1974, p. 6.

[7] Resolution No. 2990-1974 and Resolution No. 2991-1974.

[8] Telephone interview with John Shea, Office of the Attorney General, New York City, December 24, 1974.

INDEX